D1152488

Plant
Propagation
A to Z

Plant Propagation
A to Z
Growing Plants for Free

Geoff Bryant

David & Charles

A DAVID & CHARLES BOOK

First published in the UK by David & Charles Ltd in 2004
First published in New Zealand as *Growing Gardens for Free*
by David Bateman Ltd in 2003

A catalogue record for this book is available from the British Library.

ISBN 0 7153 1826 8

Printed in China by Everbest Printing Co. Ltd
for David & Charles
Brunel House   Newton Abbot   Devon

Visit our website at www.davidandcharles.co.uk

David & Charles books are available from all good bookshops; alternatively you can contact our Orderline on (0)1626 334555 or write to us at
FREEPOST EX2110, David & Charles Direct, Newton Abbot, TQ12
4ZZ (no stamp required UK mainland).

Contents

Introduction

Having a garden or even just a love of plants provides a lifetime of joy. However, that enjoyment does not always come cheaply because for all but the most esoteric designs, gardens need plants and those plants are usually bought. So it is often with economy in mind that gardeners start to propagate their own plants. However, plant propagation is even more addictive than gardening and if you have a passion for plants, propagating them provides the ultimate satisfaction. Moreover, it does not have to involve elaborate equipment and months of waiting. Hundreds of different plants can be propagated quickly with little more than a few seed trays, a good pair of pruning shears, decent potting mix and some sturdy plastic bags.

I suppose that, strictly speaking, plant propagation is a science; but for the most part it is easy. Easy, that is, once you learn the basics and break through the mystique that surrounds the idea of the "green-fingered" propagator. There is really no such creature and the expertise and equipment needed for simple plant propagation is within anyone's reach. The plants that result can be at least as good as those bought from the garden center, and they provide a sense of achievement and closeness to nature that is rare in our increasingly store-bought, manufactured age.

This book is for home gardeners whose small- to medium-scale propagation involves plenty of trial and error. While I have tried to be as precise as possible, I know that not all of my methods will work for all gardeners in all circumstances. There are usually several ways to propagate a plant, depending on the season, the available equipment and the propagator's expertise. Therefore, I have attempted to give just an outline of the general techniques involved along with highlighting some of the pitfalls to avoid. Feel free to try different methods, because a large part of being successful with plant propagation is finding out what works for you and your plants. Plenty of reading helps, too, as does developing a comprehensive understanding of plant types, families and relationships.

Finally, you need a sense of quality; don't be happy to put up with whatever plants result. Demand the same standards that you expect of a professional nursery and have no qualms about consigning your lesser efforts to the compost pile. When you start to realize that the parental pride of the propagator has to come second to producing good plants, you will know that your efforts and experiments are turning into experience and understanding.

Hardiness Zone Map

This map has been prepared to agree with a system of plant hardiness zones that have been accepted as an international standard and range from 1 to 11. It shows the minimum winter temperatures that can be expected on average in different regions.

In this book, where a zone number has been given, the number corresponds with a zone shown here. That number indicates the coldest areas in which the particular plant is likely to survive through an average winter.

Note that these are not necessarily the areas in which it will grow best. Because the zone number refers to the minimum temperatures, a plant given zone 7, for example, will obviously grow perfectly well in zone 8, but not in zone 6. Plants grown in a zone considerably higher than the zone with the minimum winter temperature in which they will survive might well grow but they are likely to behave differently. Note also that some readers may find the numbers a little conservative; we felt it best to err on the side of caution.

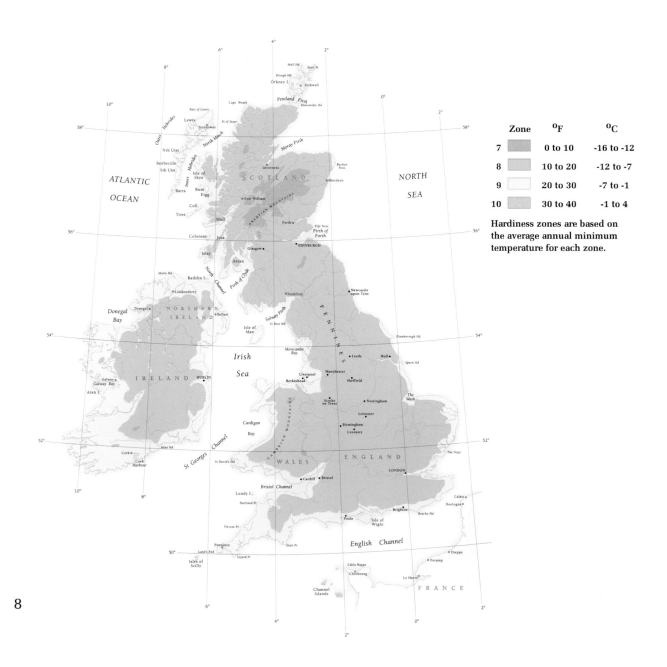

Zone	°F	°C
7	0 to 10	-16 to -12
8	10 to 20	-12 to -7
9	20 to 30	-7 to -1
10	30 to 40	-1 to 4

Hardiness zones are based on the average annual minimum temperature for each zone.

PART ONE

PLANT PROPAGATION

1
What is plant propagation?

Plant propagation is simply the technique of making two or more plants where previously there was one. By using various methods we can produce exact replicas that perpetuate much-loved varieties or we can create entirely new plants that may become favorites of the future.

Sexual reproduction and asexual (vegetative) propagation

Everyone knows that plants grow from seeds, and seeds are the most common example of sexual reproduction in plants. They are the embryos that result from fertilization of the male and female gametes and, to some extent, are directly comparable with the embryos of animals. Lower plants reproduce sexually by means of spores, which are fertilized in the presence of water; higher plants set seed through pollination. When we sow seeds or spores we are doing little more than giving nature a helping hand.

Vegetative or asexual propagation, however, is a plant phenomenon that has few direct equals in the animal world. It uses existing plant material to produce new plants and does not require the fertilization of any seed or spores. I suppose you could say that it is rather like taking a piece of skin and producing from it a new person. That kind of cloning may one day be possible, but never with the ease with which it can be done with plants. Layering, cuttings, grafting, budding and tissue culture are common methods of vegetative propagation.

Other than simply producing more plants, the main reason for propagating is to perpetuate superior forms, and the only way to be sure of doing that is to propagate them vegetatively. Because every seedling is the sum of all its parents, no seedling is exactly like its immediate parent, but shows characteristics of its entire lineage. We can breed seedling strains that are very consistent and noticeably superior to their parents, but we cannot make exact replicas by seed. Even seedlings of naturally occurring species show slight variations. Vegetative propagation answers this problem because every cutting, division, layer or the like is genetically identical to the plant from which it was taken and is therefore a perfect copy, a clone. Asexual propagation techniques may seem very clever but they are usually nothing more than sophisticated versions of things that occur naturally.

Seedheads of salsify or vegetable oyster (*Tragopogon porrifolius*). Seeds are the most common example of sexual reproduction in plants.

Low-growing, spreading plants with stems that are constantly in contact with the soil often strike roots as they spread. Layering, aerial layering and cuttings adapt this natural process to plant production. Stems that are in contact with each other for prolonged periods sometimes fuse and it is but a short step from this to grafting and budding. Even tissue culture, which demands very controlled sterile conditions, is little more than an adaptation of natural cell division.

Seed sowing is still a vital part of plant propagation, because vegetative propagation is either impractical or impossible with some plants, particularly annuals. Also, while it is theoretically possible to produce completely new plants vegetatively, by genetic manipulation, most new introductions are the result of the time-honored techniques of cross-pollination and hybridizing, the best of the resultant seedlings being perpetuated by vegetative propagation.

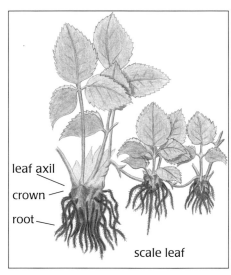

An example of natural layering.

The miracle behind vegetative propagation

How is it that plants are so easy to clone? Well, the answer is meristems or stem cells, which are clustered at the main growing points, primarily the root and stem tips. Tip or apical stem cells may develop into leaves, flowers, roots or stems depending on their position within the plant and the hormonal messages they receive. Root-tip stem cells generally develop into roots, while stem-tip meristems produce leaves, stems and flowers.

Meristematic cells also occur along the stems of plants, where their main function is to increase the diameter of the stem as the plant grows. This lateral meristem is known as the vascular cambium or cambium layer and is vitally important in plant propagation because it has the ability to produce apical meristem cells that can develop into roots and leaves. If this did not happen, our cuttings would never form roots. Also, vascular cambium can fuse with the cambium of other plants, the result being a successful graft.

These processes involve growth phytohormones known as cytokinins and auxins that initiate chemical changes. Their production and movement through the plant takes time, varying between plant types. Consequently the strike rate and strike times vary accordingly.

Stem cells occur only in the higher plants (technically those with parenchyma tissue), so algae, fungi and ferns cannot be grafted or propagated by cuttings. However, such plants are often very easily divided.

An example of vegetative reproduction: recently struck semi-ripe cuttings of *Photinia* 'Red Robin.'

Deciding which method to use

The rule-of-thumb for propagating is simple: use vegetative means when you want exact copies and choose seed only when diversity does not matter.

The exceptions, occasions when seed sowing is the preferred method, are:

1 when growing annual bedding plants and vegetables or true species;
2 when producing new hybrids;
3 when requiring large numbers of similar but not necessarily identical plants;
4 when other methods pose difficulties, such as a lack of propagation material or a reluctance to strike.

Most garden annuals and vegetables have been carefully developed and tested to come as true to type as possible from seed. Besides, they are often annuals or biennials and impossible to produce by any other means. True species usually reproduce so close to type that any variability is unlikely to be noticeable and raising them from seed maintains the genetic variability that keeps species vigorous.

In those cases conformity is fairly well guaranteed; hybridizing, on the other hand, deliberately introduces variability. Hybridizers select parent plants with special characteristics, cross them and then assess the seedlings to find those with the best characteristics of both parents and none of their vices. Coming up with a plant that approaches this ideal often takes several generations of crosses and back breeding, and once bred, the new hybrid must then be propagated vegetatively to maintain its characteristics.

Most of our modern garden perennials, shrubs and trees are hybrids or selected naturally occurring forms that cannot reproduce true to type from seed—many are infertile, too—so they must be propagated vegetatively. The same holds true for any unusual growth forms, such as sports and mutations: special flower colors or variegated foliage. To preserve them, they must be propagated vegetatively.

Various cultivars of coleus (*Solenostemon scutellariodes*). Cultivars are produced by hybridization (crossing different parent plants). They must then be propagated vegetatively to preserve their unusual characteristics.

12

There are many ways to propagate a plant vegetatively. Division, the simplest method, is most commonly used with tubers and fibrous-rooted perennials. Plants with permanent stems, and quite a few perennials, are usually raised from cuttings or by layering. Plants that are difficult to grow by other means or that would not do well on their own roots are grafted or budded. Bulbs and corms have their own special techniques or they may be grown from the numerous natural offsets they produce.

Some plants can be propagated by only one method, but most can be grown in several ways. Which method you choose largely depends on the type of plant, the material available, how many plants you require and your level of experience. Try various methods, experiment and, most importantly, learn from your mistakes. Home plant propagation is an inexact science where minute variations in propagation material, growing facilities and climate can make a vast difference to the results, so be prepared to try anything until you find what works for you.

Timing

Unless you have a particular interest in an area that requires specialized techniques, you will probably find that seed, division and cuttings make up the bulk of your propagation. While these simple techniques are easily mastered, their success or failure often depends not on how well they are done, but on when. Timing is all-important.

Many factors influence plant growth. Nurseries often use advanced climate control in their greenhouses so that it is possible to have plants ready at particular times for sale or propagation. Most of your plant propagation, however, will be with garden-sourced material and the stage

Most plants can be grown using different methods. Viburnums, such as the cranberry viburnum (*Viburnum trilobum*), can be raised from cuttings or seed.

13

of growth a plant has reached and whether or not it is a good time to attempt to propagate it will largely determine how successful you are.

Timing is most important in deciding when to divide a plant, when to take a cutting and what sort, and when to graft or bud. The precise time for each technique varies markedly depending on your local climate. In mild areas many evergreen shrubs can provide semi-ripe cuttings year round, but where winters are severe, propagation may have to stop because the plants are completely dormant and the ground is frozen solid.

Timing is less important when sowing seed because it is easy enough to use a heating pad to warm the sowing medium, though you still need to be able to keep the seedlings growing after germination.

Some people seem to know almost instinctively the right time to propagate a plant; most of us, however, need help. This book includes tables giving times for propagation, but they are only a guide, and it is a good idea to keep a diary so that you know what worked, what did not and when. Keep on experimenting, too—that seed may have germinated well in autumn, but why not sow half a packet in spring just to see what happens?

2
Propagation tools

For me, the keys to successful plant propagation are a pleasant and effective working environment, an uninterrupted flow of work and simple, functional tools.

Working environment

A good working environment is vital for your comfort and enjoyment. Particularly important is a bench at which you can stand or sit comfortably for long periods.

Presuming you do not have a conservatory attached to your house, you will also need shelter for both you and your plants. A greenhouse is not essential, at least not initially. To start with you can get by using your garage or tool-shed bench to work on, with a home-built cold frame in which to keep the plants.

Propagating frames

Cold frame

A cold frame is a large box with a glazed or clear plastic lid, usually sloping from the back to the front to allow easy access and rain run-off. It is effectively a miniature greenhouse used to harden seedlings prior to planting them out and as somewhere to keep trays of cuttings until they strike.

Cold frames have one major drawback: being small they are subject to wide variations of temperature; they heat up quickly and cool down just as fast. Because a frame would rapidly heat to very high temperatures if kept in the sun, a position in light shade or with early morning sun is best. A covering of burlap or frost-cloth will provide some additional insulation on cold winter nights but is unlikely to stop hard frosts penetrating. Adding heating cables to a frame can eliminate this problem but requires an electricity supply.

Sun frame

The sun frame is an old variation on the cold frame that was popular before mist propagation became widely available. Most suitable for

A simple cold frame for hardening off seedlings and young cuttings.

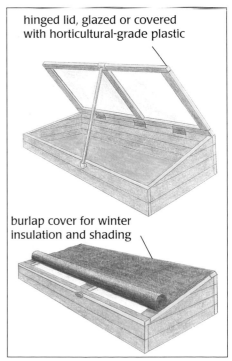

hinged lid, glazed or covered with horticultural-grade plastic

burlap cover for winter insulation and shading

hardy subjects such as conifers, it is a shallow sloping frame that is oriented to catch the maximum amount of sunlight. It has a potting soil base that allows just enough room for the cuttings to be inserted without touching the cover. The inserted cuttings are thoroughly soaked and should not require additional soil moisture until struck.

As might be expected, exposing cuttings to full sun like this will lead them to wilt, so shading is essential for the sun frame, but it should be kept to a minimum—just a light blind such as a white sheet stretched over a frame. The cuttings also need to be hand misted several times a day. The resulting combination of warmth and close humidity to some extent mimics the effects of an automated mister and bottom heat.

Indoor frame

If space is really at a premium you might consider a small self-enclosed propagating unit that can be kept indoors. These usually comprise a plastic tray with a fitted clear plastic dome-like lid, often with a heated pad on which to place the tray. Such propagators have a place in small-scale production, but with so little room for the plants to develop I think they are best restricted to seed raising only. Also, the heating pad quickly dries out the soil in the tray, which can be disastrous for young plants.

Greenhouse

Ultimately, once addicted, you will probably decide that you must have a greenhouse. Avoid the little 6 x 8 ft (1.8 x 2.4 m) greenhouses; they are really too small unless you don't mind having your work bench somewhere else. An 8 x 10 ft (2.4 x 3 m) greenhouse is about the minimum practical

Simple pipe-framed and plastic-covered greenhouse and shadehouse.

size, allowing enough room for a small bench and about 20 propagation trays with a little space left over to harden off a few plants.

Greenhouses vary in construction, functionality, durability and, of course, cost. The plastic-covered tunnel design is the cheapest and easiest to erect, but the sloping sides sometimes intrude too far into the working space. Glass greenhouses are sturdy, long-lived and usually simple to extend. While expensive and prone to breakage, they represent the best option in the long run. However, if you really cannot afford a glass greenhouse, a wooden frame with straight sides and a pitched roof covered with heavy-duty (200 micron/0.2 mm) agricultural plastic film is a good compromise. Bear in mind, though, that even the best modern plastics last only about five years before beginning to deteriorate. Rigid polycarbonate or fiberglass sheeting lasts longer, but it can be more expensive than glass.

Ventilation and insulation

Although a shadecloth cover will cool the greenhouse, adequate ventilation is vital. The rule is to have an area of venting roughly equivalent to 30 percent of the floor area of the greenhouse, so an 8 x 10 ft (2.4 x 3 m) greenhouse would require 24 sq ft (2.2 sq m) of venting.

Unless you live in a frost-free area, your greenhouse will require winter insulation. A single layer of plastic or glass will not stop a heavy frost damaging the plants. Lining the greenhouse with plastic sheeting makes a considerable difference, but the ultimate system is the purpose-built double-skinned or double-glazed greenhouse with two layers of glass or plastic separated by an air gap. This provides frost protection down to about 20°F (–7°C) for greenhouses over 800 sq ft (74 sq m), or 27°F (–3°C) for those down to 50 sq ft (5 sq m), without additional heating.

A home-built propagating house, with white shadecloth sides and roof covered with plastic film.

Shadehouse

Even if you can get by without a greenhouse, you will still need a sheltered environment to protect your tender young plants. A shadehouse provides a good intermediate level of protection for plants that are being hardened off after being moved out of a greenhouse, lessening the effects of hot sun, moisture stress and wind while also providing a little frost protection.

Modern shadehouses are usually simple wooden or pipe frames covered with woven plastic shadecloth attached to wires strung from the frame. Shadecloth is available in various densities of weave: a 32 percent shadecloth lets through two-thirds of the light and stops one-third, while a 50 percent shadecloth allows through about as much light as it stops. It is usually best to opt for the lightest cloth that provides enough protection or you may find that plants suffer from sunburn when they are eventually moved into the open garden.

Equipment

Basic needs

Your hardware needs depend on which plants you want to propagate, how many and how sophisticated or automated you want your propagation set-up to be. Raising a few seeds and divisions requires little more than some seed trays, potting mix and a sharp knife. However, you should consider the following as being necessary for a basic system that is capable of producing a wide range of plants.

Pruning shears and knives, disassembled for sharpening.

- **Pruning shears (secateurs).** Good pruning shears will last a lifetime, so don't skimp. Choose a comfortable, easily disassembled scissors style with replaceable blades. Knives are often recommended for taking cuttings, but pruning shears are far quicker, easier to use and just as effective, provided they are well maintained.
- **Two sharp knives.** First, a very sharp knife with a fine blade for budding, grafting and any other precise cutting work. Second, a general-purpose knife for chores like cutting string, and for dividing small perennials and cutting tubers. Hobby knives with snap-off replaceable blades are cheaper that fancy propagating knives and always have a razor-sharp edge.
- A good quality **spade** for breaking up clump-forming perennials, such as *Hosta* and *Phlox*. Because it will be used for levering, a steel-handled spade is preferable. Some authorities recommend using two spades or forks back to back to lever apart perennial clumps but it is a technique that I have never mastered.
- A double-sided **sharpening stone** for keeping your pruning shears, knives and spade in top condition.

- A hand-held **mister** for keeping your cuttings moist while you take them. Any small atomizer, such as an old window cleaner bottle, will do. Keep a spare handy as the nozzles tend to clog and the pump mechanisms are often short-lived.
- **Trays and pots.** Shallow trays are used for seed sowing and cuttings. Pots can be used for the same purposes, but are usually less space-efficient. The containers should be well drained and be easy to wash. You will also need a selection of small pots, sometimes called tubes, for struck cuttings and seedlings.
- A **soil sieve**, either plastic- or wooden-framed with a ¼ in (6 mm) stainless steel mesh. This is an essential item for making consistent cutting and seed-raising mixes.
- **Root-forming hormones**. These are produced naturally by the plant but adding a little extra speeds things up. While not essential, they help with hard-to-strike cuttings. They are available in powder, liquid or gel form, and which you use is largely a matter of personal preference rather than any superiority in performance.
- **Soil mixes**. You can use seed-raising, cutting and potting mixes straight from the packet, but for the sake of consistency from batch to batch I prefer to make my own. These are described later (see "Starter Mixes," p. 52).
- **Thermometers**. While not essential, thermometers are handy for keeping track of the temperature of potting mix, heated beds and your propagating area. They also confirm what any nursery worker knows: greenhouses can be really hot in summer and really cold in winter.
- **Garden chemicals**. Insecticides, fungicides and soil sterilizing agents, either organic or synthetic, have their place, though careful management is more important.
- **Water**. It may seem strange to think of water as a tool, but it is vital at every stage of propagation and its quality has a marked effect on your plants. Excessive quantities of chlorine and other salts will eventually damage plants. If you know that your local water is chemically treated, or if heavy white deposits appear on your containers, it may be wise to contact your local utility or authority for information on water quality and minerals and to find out what corrections may be required.

You will also need sundry equipment like marker pens and labels, cleaning rags, tape, wire ties and a diary to record when certain jobs were done and for keeping weather records. If you have a computer, you should consider keeping records on a database of plant types.

A mist nozzle operating. Moisture-sensor-controlled misting units are great aids to propagation, speeding up root striking and reducing labor.

Useful extra: automated mister

Other aids are useful but not essential. The one that would make the greatest difference to your cutting production is an automated mist-propagating unit. A mister keeps your cuttings moist and turgid (firm) even under the most trying conditions by ensuring that the foliage remains moist and never gets to the point of wilting. It works by using a moisture sensor to open a solenoid valve that controls nozzles delivering a fine mist. When the humidity drops, the misters are activated and they stop when the moisture level has been restored. A misting unit may seem like an extravagance but it has advantages:

- It takes the guesswork out of cutting propagation.
- It frees you from the labor of constantly tending softwood cuttings in warm weather.
- It is a good way of gently moistening fine seed after sowing.
- It can be useful in raising fern spores.

If you really need a large number of plants, even the most sophisticated system will quickly pay for itself.

Hygiene

Good hygiene is vitally important. Keep your tools and containers clean. Always wash propagating trays and pots before use and after emptying them, preferably with a disinfectant, but at least with a high pressure burst from the hose.

If you use fresh potting mix for each new tray of seeds or cuttings, soil sterilization will be unnecessary. Consign used mix to the compost pile or use it as garden mulch. It is a false economy to enhance poor soil mixes with sterilizing agents or bactericides and many of these compounds are extremely toxic. Steam is the only absolutely safe soil sterilant.

You should also regularly (every two months or so) clean your frames or greenhouse to keep down the incidence of pests and diseases. Household detergents and disinfectants (preferably biodegradable) work well, but keep them off the plants. An especially thorough annual clean-up is a good idea.

Keep an eye open for insect or disease troubles (see next chapter), and try to eliminate weed problems. It is important to prevent young plants becoming smothered by weeds. These are jobs that are easily done provided you keep on top of them, but if the weeds and pests are allowed to proliferate you may have real problems.

Prevent young plants becoming smothered by weeds. Here, freshly struck and potted camellias are being choked by chickweed.

3
Pest and disease control

Provided you always select strong, healthy propagation material and maintain hygienic growing practices, you should not be troubled by pests and diseases. Also, propagation involves keeping a fairly close watch on your plants, so if any problems do occur they are usually caught at an early stage before they get out of hand.

Young plants with very soft growth (particularly young seedlings) may be damaged by some controls, including organic sprays, and especially by overly strong oil sprays. Test any new spray on a few plants before introducing it for more general use. The safest and most effective sprays are usually the fatty acid-based insecticides and sulfur- or copper-based fungicides.

Pests

- **Aphids, thrips and whiteflies** are the most common insect pests and can be found in almost any greenhouse. They are not that difficult to control, although they breed rapidly. Pheromone traps and color lures are very good at indicating that a problem exists, but unless it is a very small infestation you will probably also need to use a mild insecticide. This is because the traps can catch only the winged adults, whereas it is the sap-sucking larvae that cause most of the damage. The larvae can be controlled with fatty acid- or oil-based sprays—complete coverage of the undersurface of the foliage, where the larvae feed, is essential.

Severe thrip damage on *Rhododendron*. Note the silvered foliage.

- **Mites** are major pests that quickly build up resistance to chemical controls. Oil sprays work by smothering the mites and suffocating them. They are effective, but good coverage is especially important as mites often lurk in inaccessible places.
- **Slugs and snails** can cause considerable damage to young seedlings and must be kept under control. Chemical baits are effective but may also poison birds and other animals, so be careful how you use them. Some of the home-made traps often mentioned in organic gardening books are worth trying.
- **Cutworms, mealy bugs, beetle larvae and weevils** are not usually

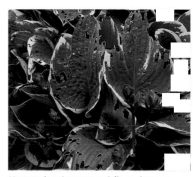

Hostas showing severe foliage damage caused by slugs and snails.

Common pests: [a] nettle leaf miner grub; [b] mealy bug on maidenhair fern; [c] adult rhododendron root weevil; [d] winged aphid.

present in sufficient numbers to cause significant damage. Cutworms are large nocturnal caterpillars that are the larvae of several species of moth. They are easily controlled by hand if you inspect the plants at night. Mealy bugs and some beetles, moths and weevils have subterranean larvae that feed on plant roots but unless their numbers are very concentrated, as sometimes happens if they occur in potted plants, they are unlikely to cause unsustainable damage. Bad infestations of all of these pests can be controlled with soil insecticides or by drenching the soil with a safe pyrethroid insecticide.

- **Sciarid flies**, the larvae of which live in the soil, are common in potting mixes and can become quite serious pests in greenhouses and other warm growing environments. Under normal conditions they have little or no effect on general plant growth, but if their larvae are present in large numbers among young seedlings they can cause considerable damage to the roots. Control them with soil insecticides.

Diseases

Fungal diseases

More young plants, especially seedlings, are damaged or killed by fungal diseases than by any other cause. This is because they are easily bruised, which allows an entry point for such diseases, and because the mild, humid propagating environment that is so ideal for young plants is also ideal for the spread of fungal problems.

The main forms of fungal disease are:

- the surface molds, such as *Botrytis* and powdery mildew, which show up as grayish deposits on the foliage or stems;
- the wilt diseases, such as *Phytophthora* and *Rhizoctonia*;
- damping-off diseases, which most commonly show up as the rotting of soft tissue.

Common rose fungal diseases. Clockwise from top left: mildew; bud with botrytis and mildew; rust; black spot.

Under greenhouse conditions fungal problems can occur at any time. Good ventilation reduces their effects by lessening the build-up of spores. Greenhouses and frames should always be well ventilated, even in winter. Uncover trays of young seedlings or struck cuttings as soon as possible. Avoid crowding your plants, as this will encourage the spread of disease. When striking cuttings under mist, make sure that the soil is not constantly waterlogged. It should be just moist enough to keep the cuttings turgid; any more will lead to rotting.

In a moist environment, systemic fungicides are undoubtedly the most effective controls, but most are unacceptable to organic gardeners. Organic sprays, such as *Melia* and *Sambucus* extracts, are useful but best used as preventives. Colloidal sulfur- and copper-based fungicides are very effective while being environmentally safe, but they may damage soft growth, so use them with care.

Rhizoctonia and *Phytophthora* wilts are soil borne and control by any chemical means, synthetic or organic, is seldom economically effective. Good drainage prevents these diseases spreading but it may also be a better idea to look for a new source of potting mix.

Controlling fungal diseases is largely a matter of good hygiene. Keep your propagating area well ventilated, use fresh potting mixes and avoid overcrowding your plants.

Viral infections

Occasionally you will also see viral diseases. These usually appear as a yellow mottling or flecking of the foliage, often with distorted new growth. Viruses are not always fatal but they are generally debilitating and nearly always incurable. Virus-infected plants should be destroyed, preferably by burning. Avoid putting them in your compost pile and most definitely do not use them for propagating.

Careful use of sprays

You can keep your use of sprays to a minimum by accurately timed preventive spraying. Rather than having to deal with pest outbreaks as they occur, it is more efficient and effective to have your spray program coinciding with crucial points in the pest's life cycle. For instance, many moths produce several generations of caterpillars in a normal growing season. Spraying at the egg-laying to hatching stage for each of these generations will give you maximum control with minimum spraying.

Nobody wants to be exposed to pesticides and there are certainly bad environmental consequences caused by their overuse. While synthetic agricultural chemicals are getting safer, newer organic controls, especially those based on naturally occurring diseases and pheromones, are now genuinely effective and are well worth trying.

Cuttings with damping-off, a common soft rot that affects the stems of seedlings and cuttings

Freshly germinated evergreen azalea seedlings.

23

Table 1: General propagation methods

The following table lists the main methods of propagation for over 1000 common genera. Individual species or cultivars may have special requirements. For example, many genera contain both annual and perennial species, and the propagation techniques vary accordingly. Refer to the later tables (seeds, division and cuttings—in Chapters 4, 5 and 6) for more details. For example, if a genus is listed as being propagated by seed and by cuttings, make sure you check both the seed and cutting lists for any specific requirements.

Certain conventions have been followed when choosing which methods to list:

- All flowering plants may be propagated from seed, so seed is listed only when it is a widely used method or when seedlings are required for producing grafting or budding stocks. For example, magnolias may take many years to flower when raised from seed, but seedling magnolias can be used as grafting stocks, so seed is listed as a method for producing magnolias.
- Division is listed for plants that produce suckers, offsets and natural layers as well as those with clumps of divisible roots or tubers.
- Virtually all plants (other than annuals) may be propagated by layering, so this is not listed separately.
- Most plants that can be grafted may also be budded, so the two methods are not differentiated.

An * in the zone column denotes an annual or a short-lived perennial that is treated as an annual in zones colder than its normal range. The zonal range is for the genus as a whole. Species within a genus vary in hardiness, so please check the hardiness of individual species before planting them outdoors.

Names in parentheses are commonly used botanical synonyms.

Plant	Seed	Divide	Cutting	Bud/Graft	Bulb	Zone
Abelia			✓			7–11
Abeliophyllum—White Forsythia			✓			5–9
Abelmoschus—Okra	✓					9–11
Abies—Fir	✓		✓	✓		4–9
Acacia—Wattle	✓		✓			8–11
Acaena—New Zealand Bur, Bidi-Bidi		✓				7–10
Acanthus—Bear's Breeches	✓	✓	✓			7–10
Acca—Feijoa	✓		✓			8–10
Acer—Maple	✓		✓	✓		4–9
Achillea—Yarrow	✓	✓				4–10

Plant	Seed	Divide	Cutting	Bud/Graft	Bulb	Zone
Ackama	✓		✓			9–11
Acmena—Lillypilly	✓		✓			9–11
Aconitum—Aconite, Monkshood	✓	✓				4–9
Acorus—Sweet Flag		✓				3–10
Actaea—Baneberry	✓	✓				3–9
Actinidia—Kiwifruit			✓	✓		5–9
Adenandra—China Flower	✓		✓			8–10
Adiantum—Maidenhair Fern		✓				8–11
Adonis		✓				5–9
Aesculus—Horse Chestnut	✓			✓		5–9
Aethionema	✓	✓	✓			7–9
Agapanthus—Lily-of-the-Nile, African Blue Lily	✓	✓				9–11
Agapetes			✓			9–11
Agastache—Anise Hyssop	✓	✓				7–10
Agathis—Kauri	✓					9–12
Agathosma—Buchu			✓			9–10
Agave	✓	✓				8–12
Ageratina	✓	✓	✓			6–10
Ageratum—Floss Flower	✓		✓			9–12
Agonis—Willow Myrtle	✓		✓	✓		9–10
Agrostis—Bent Grass	✓	✓				5–10
Ailanthus—Tree of Heaven	✓		✓			6–10
Ajuga—Bugle		✓	✓			5–10
Akebia	✓		✓			5–9
Albizia—Silk Tree	✓		✓	✓		8–12
Alcea—Hollyhock	✓	✓				4–10
Alchemilla—Lady's Mantle	✓	✓				4–9
Alectryon—Titoki, New Zealand Oak	✓					9–11
Allamanda			✓			11–12
Allium—Onion	✓	✓			✓	6–10
Alnus—Alder	✓			✓		6–9
Aloe	✓	✓	✓			9–11
Alonsoa	✓		✓			9–10
Alopecurus—Foxtail Grass	✓	✓				3–10
Aloysia (*Lippia*)—Lemon-scented Verbena	✓		✓			8–11
Alpinia—Ornamental Ginger		✓				10–12
Alseuosmia			✓			9–10
Alstroemeria—Peruvian Lily	✓	✓				7–10
Alyssum	✓	✓	✓			7–10
Amaranthus—Love-Lies-Bleeding	✓					10–12/*
Amaryllis—Belladonna Lily	✓				✓	8–11
Amelanchier—Juneberry, Serviceberry, Shadbush	✓		✓			4–9
Ammi—Bishop's Weed	✓					6–10
Ampelopsis			✓			4–9
Anacyclus		✓				6–10
Anagallis—Pimpernel	✓	✓				7–10
Anaphalis—Pearly Everlasting	✓					5–9
Anchusa—Alkanet	✓	✓	✓			5–10

Plant	Seed	Divide	Cutting	Bud/Graft	Bulb	Zone
Andromeda—Bog Rosemary	✓		✓			2–9
Androsace—Rock Jasmine	✓		✓			5–9
Anemone—Windflower	✓	✓			✓	6–10
Anemonella—Rue Anemone	✓	✓				4–9
Anethum—Dill	✓					5–10
Angelica	✓					5–9
Angophora—Apple Gum	✓					9–11
Anigozanthus—Kangaroo Paw	✓	✓				9–11
Annona—Cherimoya	✓			✓		10–12
Anomotheca					✓	8–11
Anthemis	✓	✓	✓			4–10
Anthericum—Spider Plant	✓	✓				7–10
Anthriscus—Chervil, Cow Parsley	✓					5–10
Antigonon—Coral Vine	✓		✓			9–11
Antirrhinum—Snapdragon	✓		✓			6–10/*
Apium—Celery, Celeriac	✓					5–10
Aptenia	✓	✓	✓			9–11
Aquilegia—Columbine	✓	✓				3–10
Arabis—Rock Cress	✓	✓				5–10
Araucaria	✓					8–12
Araujia	✓		✓			9–11
Arbutus—Strawberry Tree	✓		✓			7–10
Archeria	✓					8–10
Arctostaphylos—Bearberry, Manzanita			✓			4–10
Arctotheca—Cape Weed	✓	✓				9–11
Arctotis—African Daisy	✓	✓	✓			9–11
Ardisia—Marlberry			✓			9–11
Arenaria—Sandwort	✓	✓				6–10
Arisaema—Jack-in-the Pulpit	✓	✓			✓	6–9
Arisarum—Mouse Plant	✓	✓			✓	6–10
Aristea		✓				8–11
Aristolochia—Dutchman's Pipe, Birthwort	✓	✓	✓			5–12
Aristotelia—Wineberry			✓			8–10
Armeria—Thrift, Sea Pink	✓	✓				5–9
Armoracia—Horseradish		✓				5–10
Arnica	✓					4–9
Aronia—Chokeberry			✓			4–9
Artemisia—Wormwood, Mugwort	✓	✓	✓			4–10
Arthropodium—Renga Renga Lily	✓	✓				8–11
Arum (not Zantedeschia)	✓	✓				7–10
Aruncus—Goat's Beard	✓	✓				4–9
Arundinaria—Bamboo		✓				6–10
Arundo—Giant Reed		✓				8–11
Asarina (Maurandya)—Twining Snapdragon	✓					7–10
Asclepias—Milkweed, Swan Plant	✓		✓			4–11
Asparagus	✓	✓				4–9
Asphodeline—Jacob's Rod	✓	✓				6–10
Aspidistra—Cast-iron Plant		✓				8–11

Plant	Seed	Divide	Cutting	Bud/Graft	Bulb	Zone
Asplenium—Spleenwort		✓				5–12
Astelia	✓	✓				8–10
Aster	✓	✓	✓			3–10
Astilbe—False Spiraea	✓	✓				6–10
Astrantia—Masterwort	✓	✓				5–9
Athyrium—Lady Fern		✓				5–10
Atriplex—Salt Bush			✓			6–10
Aubrieta—Rock Cress	✓	✓	✓			4–9
Aucuba—Japanese Laurel			✓			7–10
Aulax	✓		✓			9–11
Aurinia—Yellow Alyssum		✓				4–10
Azara—Vanilla Tree	✓		✓			8–10
Azolla		✓				7–12
Azorella		✓	✓			8–10
Babiana—Baboon Flower	✓				✓	9–11
Backhousia			✓			9–11
Baeckia	✓		✓			9–11
Bambusa—Bamboo		✓				9–12
Banksia	✓		✓			8–11
Baptisia—False Indigo	✓	✓	✓			4–10
Bassia (*Kochia*)	✓					*
Bauera			✓			9–10
Bauhinia—Orchid Tree			✓			10–12
Beaufortia	✓		✓			9–11
Beaumontia—Herald's Trumpet	✓		✓			9–12
Begonia	✓	✓	✓			9–12/*
Beilschmiedia—Iawa	✓					9–11
Belamcanda—Leopard Lily	✓	✓				8–11
Bellis—Daisy	✓	✓	✓			3–10
Berberidopsis—Coral Vine	✓		✓			8–10
Berberis—Barberry	✓		✓			5–10
Bergenia—Pigsqueak, Elephant's Ear	✓	✓				4–9
Berzelia	✓		✓			9–11
Beschorneria		✓				9–12
Beta—Garden Beets	✓					5–10
Betula—Birch	✓			✓		2–9
Bignonia—Cross-vine			✓			6–10
Billardiera—Apple Berry			✓			8–10
Billbergia—Vase Plant		✓				10–12
Blandfordia—Christmas Bells	✓	✓				9–11
Blechnum		✓				7–11
Bletilla—Chinese Ground Orchid		✓				7–10
Boltonia—False Chamomile	✓	✓	✓			4–9
Bomarea		✓				8–11
Borago—Borage	✓					5–10
Boronia			✓			9–11
Bougainvillea	✓		✓			10–12
Bouvardia			✓			10–11

Plant	Seed	Divide	Cutting	Bud/Graft	Bulb	Zone
Brachychiton—Kurrajong	✓			✓		9–12
Brachyscome—Swan River Daisy	✓		✓			9–11/*
Brachyglottis			✓			7–10
Brachysema—Swan River Pea			✓			9–11
Brassica crops	✓					5–10
Briza—Quaking Grass, Snakegrass	✓	✓				5–11
Brodiaea	✓				✓	8–10
Bromeliads		✓				8–12
Browallia—Bush Violet	✓					9–11
Brugmansia (*Datura*)—Angel's Trumpet			✓			9–12
Brunfelsia—Yesterday, Today and Tomorrow			✓			9–12
Brunnera—Siberian Bugloss		✓				3–9
Brunsvigia					✓	9–11
Buddleja	✓		✓			7–10
Bulbinella	✓	✓				8–10
Bupthalmum—Ox Eye	✓	✓	✓			4–9
Butia—Yatay Palm, Jelly Palm	✓					9–11
Butomus—Flowering Rush	✓	✓				5–9
Buxus—Box			✓			6–10
Caesalpinia	✓		✓			10–12
Calamintha—Calamint	✓	✓	✓			4–10
Calandrinia—Rock Purslane	✓					8–12
Calceolaria—Slipper Flower	✓		✓			6–10/*
Calendula	✓					6–10
Calla—Water Arum, Bog Arum		✓				2–9
Calliandra—Powderpuff Tree			✓			9–12
Callicarpa—Beautyberry	✓		✓			6–12
Callistemon—Bottlebrush	✓		✓			8–11
Callistephus—China Aster	✓					*
Callitris—Cypress Pine	✓					9–11
Calluna—Heather	✓	✓	✓			4–9
Calocedrus—Incense Cedar	✓		✓			5–9
Calocephalus			✓			9–10
Calochortus—Mariposa Tulip	✓				✓	5–9
Calodendrum—Cape Chestnut	✓			✓		9–11
Calothamnus—Net Bush			✓			9–11
Caltha		✓				3–9
Calycanthus—Allspice	✓					6–10
Calystegia—Bindweed	✓	✓	✓			5–10
Calytrix—Fringe Myrtle			✓			9–11
Camassia—Quamash	✓				✓	4–9
Camellia	✓		✓			6–10
Campanula—Bellflower	✓	✓	✓			3–10
Campsis—Trumpet Vine	✓		✓			6–10
Canarina—Canary Bellflower		✓				9–11
Canna—Indian Shot	✓	✓				9–12
Cantua—Sacred Flower of the Incas			✓			9–11
Capsicum—Pepper	✓					9–12/*

Plant	Seed	Divide	Cutting	Bud/Graft	Bulb	Zone
Caragana—Peashrub	✓			✓		3–9
Cardamine—Bittercress	✓	✓	✓			4–9
Cardiocrinum—Giant Lily	✓				✓	7–9
Carex—Sedge		✓				3–9
Carica—Papaya	✓		✓			10–12
Carissa—Natal Plum	✓		✓			9–12
Carmichaelia	✓		✓			8–10
Carpenteria—Tree Anemone			✓			8–10
Carpinus—Hornbeam	✓			✓		5–9
Carpodetus	✓					8–10
Carthamnus—Safflower	✓					7–11
Carum—Caraway	✓					5–10
Carya—Hickory, Pecan	✓			✓		4–10
Caryopteris—Bluebeard			✓			6–10
Casimiroa—Sapote	✓			✓		9–12
Cassia (not *Senna*)	✓		✓			10–12
Cassinia	✓		✓			8–11
Cassiope		✓	✓			2–9
Castanea—Chestnut	✓			✓		4–9
Castanospermum—Moreton Bay Chestnut	✓					10–12
Casuarina—She Oak, Horsetail Tree	✓		✓			9–12
Catalpa—Indian Bean	✓		✓	✓		4–10
Catananche—Cupid's Dart	✓	✓	✓			7–10
Cattleya		✓				10–12
Cavendishia			✓			9–10
Ceanothus—California Lilac			✓			7–10
Cedronella—Balm of Gilead	✓		✓			8–11
Cedrus—Cedar	✓		✓	✓		6–10
Celastrus—Bittersweet			✓			4–9
Celmisia—Mountain Daisy	✓	✓				6–9
Celosia—Cockscomb	✓					*
Celtis—Hackberry			✓			4–10
Centaurea—Cornflower, Knapweed	✓	✓	✓			4–10
Cephalaria		✓				3–9
Cerastium	✓	✓				3–10
Ceratonia—Carob	✓					9–11
Ceratopetalum	✓		✓			9–11
Ceratostigma—Plumbago	✓		✓			6–10
Cercidiphyllum—Katsura Tree			✓			6–9
Cercis—Redbud, Judas Tree	✓					5–9
Ceropegia			✓			9–12
Cestrum	✓		✓			9–12
Chaenomeles—Flowering Quince, Japonica	✓		✓	✓		5–10
Chamaecyparis—False Cypress			✓			4–10
Chamaedorea	✓					10–12
Chamaemelum—Chamomile	✓	✓	✓			5–10
Chamelaucium			✓			10–11
Chimonanthus—Wintersweet	✓					6–10

Plant	Seed	Divide	Cutting	Bud/Graft	Bulb	Zone
Chionanthus—Fringe Tree	✓					6–10
Chionodoxa—Glory of the Snow	✓				✓	4–9
Chlidanthus—Sea Daffodil					✓	9–11
Chlorophytum—Spider Plant		✓				9–11
Choisya—Mexican Orange Blossom			✓			8–11
Chordospartium	✓					8–10
Chorizema—Flame Pea	✓		✓			9–11
Chrysanthemum	✓	✓	✓			7–10
Cichorium—Chicory, Endive	✓					4–10
Cimicifuga—Bugbane		✓				3–9
Cinnamomum—Camphor Tree, Cinnamon			✓			9–12
Cissus—Kangaroo Vine			✓			10–12
Cistus—Rock Rose			✓			8–10
Citrullus—Watermelon	✓					10–12/*
Citrus—Lemon, Orange, Grapefruit, etc.	✓		✓	✓		9–12
Cladanthus (*Anthemis*)	✓					*
Clarkia/Godetia	✓					*
Claytonia—Spring Beauty	✓	✓				4–8
Clematis—Virgin's Bower	✓	✓	✓			4–10
Cleome—Spider Flower	✓					9–11/*
Clerodendrum—Glory Bower, Glory Flower, Butterfly Bush	✓		✓			8–12
Clethra—Lily of the Valley Tree, Summersweet			✓			5–10
Clianthus—Kaka Beak, Parrot's Bill	✓		✓			8–10
Clivia—Kaffir Lily	✓		✓			10–11
Clytostoma			✓			9–12
Cobaea—Cup and Saucer Vine	✓		✓			9–11
Colchicum—Autumn Crocus	✓	✓			✓	6–9
Coleonema			✓			9–10
Colquhounia			✓			8–10
Colutea			✓			6–10
Congea			✓			10–12
Consolida—Larkspur	✓					*
Convallaria—Lily-of-the-Valley		✓				3–9
Convolvulus—Bindweed			✓			5–10
Coprosma	✓		✓			8–10
Corallospartium	✓		✓			8–10
Cordyline—Ti, Cabbage Tree	✓		✓			8–12
Coreopsis—Tickseed	✓	✓	✓			4–10
Coriandrum—Coriander	✓					*
Cornus—Dogwood	✓		✓			3–9
Corokia	✓		✓			8–10
Coronilla—Crown Vetch	✓	✓	✓			6–10
Correa—Australian Fuchsia			✓			8–10
Cortaderia—Pampas Grass, Toe Toe	✓	✓				6–10
Corydalis		✓				6–10
Corylopsis—Winter Hazel	✓		✓			6–9
Corylus—Hazel, Filbert	✓		✓	✓		4–9
Corynocarpus—Karaka	✓		✓			9–11

Plant	Seed	Divide	Cutting	Bud/Graft	Bulb	Zone
Cosmos—Mexican Aster	✓	✓	✓			8–11/*
Cotinus—Smoke Bush			✓			5–10
Cotoneaster	✓		✓			4–10
Cotula	✓	✓				8–10
Crassula			✓			9–11
Crataegus—Hawthorn, May	✓		✓	✓		4–10
Crinodendron			✓			8–10
Crinum	✓	✓			✓	7–12
Crocosmia—Montbretia		✓			✓	8–11
Crocus		✓			✓	4–9
Crotalaria—Rattlebox			✓			9–12
Cryptomeria Japanese Cedar	✓		✓			7–10
Cucumis—Cucumber/Melon	✓					10–12/*
Cucurbita—Pumpkin/Squash	✓					10–12/*
Cuphea	✓		✓			9–12
Cupressocyparis—Leyland Cypress			✓			5–10
Cupressus—Cypress	✓		✓	✓		7–11
Cyananthus	✓		✓			4–9
Cyathodes	✓		✓			8–11
Cyclamen	✓					6–10
Cydonia—Quince	✓			✓		6–9
Cymbalaria—Kenilworth Ivy	✓	✓	✓			4–10
Cymbidium		✓				9–11
Cymbopogon—Lemon Grass	✓	✓				10–12
Cynara—Cardoon, Globe Artichoke	✓	✓				6–10
Cynodon—Bermuda Grass		✓				9–12
Cynoglossus	✓					5–9/*
Cypella	✓				✓	8–10
Cyperus—Umbrella Plant, Papyrus	✓	✓				9–12
Cyphomandra—Tamarillo	✓		✓			9–11
Cypripedium—Slipper Orchid		✓				4–9
Cyrtanthus—Fire Lily, Scarborough Lily	✓	✓			✓	9–11
Cytisus—Broom	✓		✓			6–10
Daboecia—Irish Heath	✓	✓	✓			7–9
Dacrydium—Rimu	✓		✓			8–10
Dactylorhiza—Marsh Orchid		✓				4–10
Dahlia	✓	✓	✓			8–10
Dais—Pompon Bush	✓					9–11
Dampiera			✓			9–10
Daphne			✓			5–10
Daucus—Carrot	✓					4–10/*
Davallia—Hare's Foot Fern		✓				9–12
Davidia—Dove Tree, Handkerchief Tree	✓		✓			7–9
Decaisnea	✓					5–9
Delphinium	✓	✓	✓			3–10
Desfontainea	✓		✓			8–9
Deutzia			✓			4–10
Dianella—Flax Lily		✓				8–11

Plant	Seed	Divide	Cutting	Bud/Graft	Bulb	Zone
Dianthus—Pink, Carnation	✓		✓			3–10
Diascia—Twinspur	✓		✓			8–10
Dicentra—Bleeding Heart	✓	✓	✓			3–9
Dichondra	✓	✓				9–11
Dictamnus—Burning Bush	✓	✓	✓			3–9
Dierama—Wand Flower	✓	✓			✓	8–10
Dietes syn *Moraea*—Fortnight Lily, Wild Iris		✓				9–11
Digitalis—Foxglove	✓	✓	✓			4–10
Dimorphotheca—African Daisy	✓		✓			8–10/*
Dioscorea—Yam		✓				10–12
Diosma—Breath of Heaven			✓			9–10
Diospyros—Persimmon, Sapote	✓		✓	✓		5–12
Dipelta				✓		6–9
Dipsacus—Teasel	✓					4–10
Disanthus			✓			8–10
Distictis (*Phaedranthus*)			✓			9–11
Dodecatheon—Shooting Star	✓	✓				3–9
Dodonea—Hop Bush, Ake Ake	✓		✓			9–12
Dombeya—Wedding Flower	✓		✓			9–12
Doodia—Rasp Fern		✓				9–11
Doronicum—Leopard's Bane	✓	✓				5–9
Dorotheanthus—Livingstone Daisy	✓					*
Draba	✓	✓				4–9
Dracaena—Dragon Tree			✓			9–12
Dracocephalum—Dragon's Head		✓				3–9
Dracophyllum—Dragon's Head	✓		✓			3–9
Dracunculus—Dragon Arum		✓				9–11
Dregea			✓			9–11
Drimys—Winter's Bark			✓			8–10
Drosanthemum			✓			9–11
Dryas—Mountain Avens	✓		✓			2–9
Dysoxylum	✓		✓			9–12
Eccremocarpus—Glory Flower	✓		✓			8–10
Echeveria	✓	✓				8–11
Echinacea—Cone Flower	✓	✓	✓			3–10
Echinops—Globe Thistle	✓	✓	✓			3–10
Echinopsis—Sea Urchin Cactus		✓				9–11
Echium	✓		✓			7–10
Edgeworthia—Paper Bush			✓			8–10
Elaeagnus—Oleaster	✓		✓			3–10
Elaeocarpus	✓		✓			9–12
Embothrium—Chilean Fire Bush			✓			8–10
Enkianthus			✓			6–10
Entelea—Whau	✓		✓			9–11
Epacris	✓		✓			9–11
Ephedra—Mormon Tea, Joint Fir	✓					4–10
Epidendrum		✓				10–12
Epilobium (incl. *Zauschneria*)—Willow Herb		✓	✓			3–10

Plant	Seed	Divide	Cutting	Bud/Graft	Bulb	Zone
Epimedium—Bishop's Hat, Barrenwort		✓				5–9
Episcia			✓			10–12
Equisetum—Horsetail		✓				2–9
Eranthis—Winter Aconite	✓	✓			✓	4–9
Eremurus—Foxtail Lily	✓	✓				4–9
Erica—Heath	✓	✓	✓			5–10
Erigeron—Fleabane	✓		✓			3–11
Erinus—Alpine Balsam	✓					6–9
Eriobotrya—Loquat	✓		✓	✓		8–11
Eriocephalus—Kapok Bush			✓			9–11
Eriogonum—Wild Buckwheat	✓					5–11
Eriophorum—Cotton Grass		✓				4–9
Eriostemon—Waxflower			✓			9–11
Erodium—Stork's Bill	✓	✓	✓			6–10
Eruca—Sweet Rocket	✓					⋆
Eryngium—Sea Holly	✓		✓			4–10
Erysimum—Wallflower	✓		✓			4–10/⋆
Erythrina—Coral Tree	✓		✓			9–12
Erythronium—Dog's-tooth Violet	✓	✓			✓	3–9
Escallonia			✓			8–10
Eschscholzia—California Poppy	✓					7–10/⋆
Eucalyptus—Gum Tree	✓					8–12
Eucomis—Pineapple Lily	✓	✓			✓	8–10
Eucryphia—Roble de Chile, Pinkwood			✓			8–10
Eugenia	✓		✓			10–12
Euonymus—Spindle Tree			✓			5–10
Eupatorium	✓	✓	✓			5–11
Euphorbia	✓		✓			5–11
Euryops			✓			8–11
Eustoma (*Lisianthius*)	✓					9–11/⋆
Eutaxia			✓			9–10
Exacum	✓					10–12/⋆
Exochorda—Pearlbush	✓		✓			6–9
Fabiana			✓			8–10
Fagopyrum—Buckwheat	✓					3–9
Fagus—Beech	✓			✓		4–10
Fallopia—Lace Vine, Fleece Flower	✓	✓	✓			4–10
Farfugium		✓				5–9
Fatshedera			✓			7–11
Fatsia—Japanese Aralia	✓		✓			8–11
Felicia—Kingfisher Daisy	✓		✓			9–11/⋆
Ficus—Fig	✓		✓			8–12
Filipendula—Meadowsweet	✓	✓				3–9
Foeniculum—Fennel	✓					5–10
Forsythia			✓			4–9
Fothergilla			✓			5–10
Fragaria—Strawberry	✓	✓				4–10
Francoa—Bridal Wreath	✓	✓	✓			7–10

Plant	Seed	Divide	Cutting	Bud/Graft	Bulb	Zone
Franklinia	✓		✓			7–10
Fraxinus—Ash	✓			✓		4–10
Freesia	✓	✓			✓	9–10
Fremontodendron—Flannel Bush	✓					8–10
Fritillaria—Fritillary	✓	✓			✓	4–10
Fuchsia	✓		✓			7–11
Gaillardia—Blanket Flower	✓	✓	✓			5–10/*
Galanthus—Snowdrop	✓				✓	4–9
Galax—Beetleweed, Wandflower		✓				4–9
Galega—Goat's Rue	✓	✓				5–10
Galium—Bedstraw, Woodruff		✓				4–10
Galphimia			✓			9–11
Galtonia—Summer Hyacinth	✓				✓	6–10
Gardenia	✓		✓			9–11
Garrya—Tassel Tree, Silk-tassel Bush			✓			8–11
Gaultheria—Wintergreen, Snowberry	✓	✓	✓			4–9
Gaura		✓	✓			5–10
Gazania—Treasure Flower	✓	✓	✓			9–11
Geissorhiza	✓				✓	8–10
Gelsemium—Carolina Jessamine			✓			8–10
Geniostoma—New Zealand Privet	✓		✓			9–10
Genista—Broom			✓			4–10
Gentiana—Gentian	✓	✓	✓			4–9
Geranium	✓	✓	✓			4–10
Gerbera—Transvaal Daisy	✓	✓				9–11
Geum—Avens	✓	✓	✓			3–9
Ginkgo—Maidenhair Tree	✓		✓			4–10
Gladiolus	✓				✓	6–10
Glaucium—Horned Poppy	✓					7–10
Glechoma		✓	✓			6–10
Gleditsia—Honeylocust	✓		✓	✓		4–10
Globularia—Globe Daisy	✓	✓				5–9
Gloriosa—Glory Lily, Flame Lily	✓	✓			✓	10–12
Glycyrrhiza—Liquorice		✓				8–10
Gomphrena	✓					*
Goodia	✓		✓			9–11
Gordonia	✓		✓			6–11
Grevillea—Spider Flower	✓		✓			8–12
Grewia			✓			9–11
Greyia			✓			9–11
Griselinea			✓			8–10
Gunnera	✓	✓				7–10
Gymnocladus—Kentucky Coffee Tree	✓					4–10
Gypsophila—Baby's Breath	✓	✓	✓			4–10
Haemanthus—Cape Tulip, Blood Lily	✓	✓			✓	9–11
Hakea	✓		✓			8–11
Halesia—Silverbell, Snowdrop Tree	✓					3–9
Halimiocistus			✓			8–10

Plant	Seed	Divide	Cutting	Bud/Graft	Bulb	Zone
Halimium—Sun Rose			✓			8–9
Hamamelis—Witch Hazel	✓		✓			4–9
Hardenbergia—Coral Pea	✓		✓			9–11
Harpephyllum—Kaffir Plum	✓					10–11
Haworthia		✓				9–11
Hebe	✓		•✓			7–10
Hedera—Ivy		✓	✓			5–11
Hedycarya—Pigeon Wood	✓		✓			10–11
Hedychium—Ginger Lily	✓	✓				9–12
Helenium—Sneezeweed	✓	✓	✓			4–10
Helianthemum—Sun Rose, Rock Rose	✓		✓			6–10
Helianthus—Sunflower	✓	✓	✓			4–9
Helichrysum—Everlasting Daisy	✓	✓	✓			7–11
Helictotrichon—Oatgrass		✓				4–9
Heliophila—Cape Stock	✓					9–10
Heliopsis—Orange Sunflower, Ox Eye	✓	✓	✓			4–9
Heliotropium—Heliotrope, Cherry Pie	✓		✓			9–11/*
Helipterum—Strawflower	✓					9–11/*
Helleborus—Hellebore, Lenten Rose	✓	✓				5–9
Hemerocallis—Daylily	✓	✓			✓	4–10
Hepatica—Liverleaf		✓				5–9
Hermodactylus—Snake's Head Iris	✓	✓				7–10
Herniara		✓				5–9
Herpolirion	✓	✓				9–10
Hesperantha					✓	9–10
Hesperis—Damask Violet	✓					3–9
Heterocentron (Heeria)—Spanish Shawl		✓	✓			10–11
Heuchera—Coral Bells	✓	✓				4–10
Hibbertia—Guinea Gold Vine			✓			9–11
Hibiscus	✓	✓	✓			7–12
Hippeastrum (Amaryllis)	✓				✓	10–11
Hippocrepis—Horseshoe Vetch		✓				*
Hippophae—Buckthorn	✓		✓			2–9
Hoheria—Lacebark	✓		✓			8–10
Holmskioldia			✓			10–11
Hosta—Plantain Lily	✓	✓				6–10
Hovea—Purple Pea			✓			9–11
Hovenia—Raisin Tree			✓			8–10
Hoya—Wax Flower	✓		✓			10–11
Humulus—Hops	✓	✓	✓			5–9
Hyacinthoides—Bluebell	✓	✓			✓	5–9
Hyacinthus—Hyacinth	✓				✓	5–9
Hydrangea			✓			6–10
Hymenanthera	✓					8–11
Hymenosporum—Australian Frangipani	✓		✓			9–11
Hypericum—St. John's Wort	✓	✓	✓			6–10
Hypoestes—Freckle Face, Polka-dot Plant	✓		✓			10–12/*
Hyssopus—Hyssop	✓	✓	✓			3–1

Plant	Seed	Divide	Cutting	Bud/Graft	Bulb	Zone
Iberis—Candytuft	✓	✓	✓			5–11
Idesia—Wonder Tree	✓		✓			6–10
Ilex—Holly	✓		✓			5–10
Impatiens—Busy Lizzie, Balsam	✓		✓			7–12
Incarvillea	✓	✓				6–10
Indigofera—Indigo	✓	✓	✓			7–11
Inula	✓	✓				5–10
Iochroma			✓			9–12
Ipheion—Spring Starflower		✓				6–10
Ipomoea (incl. *Mina*)—Morning Glory	✓	✓	✓			8–12
Iris	✓	✓			✓	4–10
Isoplexis			✓			9–11
Isopogon	✓					9–11
Itea—Sweetspire			✓			5–10
Ixia	✓				✓	9–10
Ixiolirion	✓				✓	7–10
Jacaranda	✓		✓			9–11
Jasione—Sheep's Bit	✓					5–10
Jasminum—Jasmine	✓		✓			7–12
Jeffersonia—Twinleaf	✓	✓				5–9
Jovellana	✓		✓			9–11
Juglans—Walnut	✓		✓	✓		4–10
Juncus—Rush		✓				4–10
Juniperus—Juniper			✓			4–10
Justicia (*Beloperone, Jacobinia*)	✓		✓			9–12
Kadsura			✓			7–11
Kalanchoe	✓		✓			9–12
Kalmia—Mountain Laurel, Sheep Laurel	✓		✓			3–9
Kalmiopsis	✓		✓			7–9
Kerria			✓			5–10
Knautia	✓		✓			6–10
Knightia—NZ Honeysuckle, Rewa Rewa	✓					9–10
Kniphofia—Red-hot Poker	✓	✓				7–10
Koelreuteria—Golden Rain-tree	✓		✓			5–12
Kolkwitzia—Beautybush			✓			4–9
Kunzea	✓		✓			8–11
Laburnum—Golden Chain Tree	✓		✓	✓		3–9
Lachenalia—Cape Cowslip	✓				✓	9–11
Lactuca—Lettuce	✓					7–12/*
Lagerstroemia—Crape Myrtle	✓		✓			6–11
Lagunaria—Norfolk Island Hibiscus			✓			10–11
Lambertia			✓			9–11
Lamium—Dead Nettle		✓	✓			6–10
Lampranthus—Ice Plant			✓			9–11
Lantana	✓		✓			9–12
Lapageria—Chilean Bellflower	✓					9–10
Lapeirousia	✓				✓	9–12
Larix—Larch	✓			✓		2–9

Plant	Seed	Divide	Cutting	Bud/Graft	Bulb	Zone
Lathyrus—Sweet Pea, Wild Pea	✓	✓	✓			5–10/*
Laurelia	✓		✓			9–11
Laurentia		✓				6–10
Laurus—Bay Laurel			✓			8–11
Lavandula—Lavender	✓		✓			8–10
Lavatera—Tree Mallow	✓		✓			6–11/*
Ledum—Labrador Tea			✓			2–8
Leonotis—Lion's Ear		✓	✓			9–11
Leontopodium—Edelweiss	✓	✓				4–9
Leonurus—Motherwort	✓	✓				3–9
Leptospermum—Tea Tree, Manuka	✓		✓			8–11
Leschenaultia	✓		✓			9–11
Lespedeza			✓			6–10
Leucadendron	✓		✓			9–11
Leucocoryne					✓	9–11
Leucojum—Snowflake	✓				✓	4–10
Leucopogon	✓		✓			9–10
Leucospermum	✓		✓			9–11
Leucothoe	✓		✓			5–10
Levisticum—Lovage	✓		✓			4–10
Lewisia	✓					4–10
Liatris—Gayfeather, Snakeroot	✓	✓	✓			3–10
Libertia	✓	✓				8–10
Libocedrus	✓		✓			8–10
Ligularia—Leopard Plant		✓				4–9
Ligustrum—Privet			✓			6–11
Lilium—Lily	✓				✓	5–10
Limnanthes—Meadow Foam	✓					8–10
Limonium (*Statice*)	✓	✓				5–10
Linaria—Toadflax	✓	✓				5–10
Lindera—Spice Bush	✓		✓			5–10
Linnaea—Twinflower		✓				2–8
Linum—Flax	✓	✓	✓			5–10
Liquidambar—Sweetgum	✓		✓	✓		5–10
Liriodendron—Tulip Tree	✓			✓		5–10
Liriope—Lily Turf	✓	✓				4–10
Lithodora (*Lithospermum*)			✓			7–10
Lithops—Living Stones	✓					9–11
Littonia—Climbing Lily	✓	✓				9–11
Lobelia	✓	✓	✓			7–11/*
Lobularia—Sweet Alyssum	✓					*
Lomatia			✓			8–10
Lonicera—Honeysuckle			✓			5–11
Lophomyrtus (*Myrtus*)			✓			8–10
Loropetalum—Fringe Flower			✓			9–11
Lotus		✓	✓			6–11
Luculia	✓		✓			9–11
Lunaria—Honesty	✓					8–10

Plant	Seed	Divide	Cutting	Bud/Graft	Bulb	Zone
Lupinus—Lupin(e)	✓	✓	✓			4–10
Lychnis—Campion, Catchfly	✓	✓				3–10
Lycopersicon—Tomato	✓					10–12/*
Lycoris—Spider Lily	✓				✓	7–10
Lyonia—Huckleberry, Fetterbush			✓			5–9
Lysichiton—Skunk Cabbage	✓	✓				5–9
Lysimachia—Loosestrife		✓	✓			4–10
Lythrum—Loosestrife	✓	✓	✓			3–10
Macadamia	✓		✓	✓		9–11
Macfadyena (*Doxantha*)—Cat's Claw Vine			✓			9–12
Macleaya—Plume Poppy		✓	✓			4–10
Magnolia	✓		✓	✓		4–10
Mahonia	✓		✓			5–11
Malcolmia	✓					8–11
Malus—Apple, Crabapple	✓		✓	✓		3–10
Malva—Mallow	✓		✓			3–10
Mammillaria—Pincushion Cactus	✓	✓				9–11
Mandevilla	✓		✓			9–12
Manettia—Firecracker			✓			10–11
Marianthus			✓			8–11
Marrubium—Horehound	✓	✓	✓			4–10
Matricaria	✓					6–10
Matthiola—Stock	✓					6–10
Maytenus—Mayten			✓			8–10
Mazus		✓	✓			4–10
Meconopsis	✓					6–10
Melaleuca—Paperbark	✓		✓			9–12
Melia—Bead Tree			✓			8–12
Melicope			✓			10–12
Melicytus	✓		✓			9–11
Melissa—Balm	✓	✓	✓			4–10
Mentha—Mint	✓	✓	✓			4–10
Menziesia—Minnie Bush, Fool's Huckleberry	✓		✓			5–10
Mertensia		✓				3–10
Meryta—Puka	✓		✓			10–12
Mesembryanthemum—Ice Plant	✓		✓			9–11
Mespilus—Medlar	✓			✓		4–9
Metasequoia—Dawn Redwood	✓		✓			5–10
Metrosideros—Pohutukawa, Rata	✓		✓			8–11
Michaelia			✓			9–11
Micromyrtus			✓			9–10
Microsorium		✓				9–12
Mimetes			✓			9–10
Mimulus—Musk, Monkey Flower	✓	✓	✓			6–11
Mirabilis—Umbrella Wort, Marvel of Peru	✓	✓				8–11
Mirbelia			✓			9–11
Mitraria			✓			9–11
Moluccella—Bells of Ireland	✓					7–10

Plant	Seed	Divide	Cutting	Bud/Graft	Bulb	Zone
Monarda—Bergamot	✓	✓	✓			4–10
Monstera—Fruit Salad Plant	✓		✓			10–12
Moraea—Fortnight Lily					✓	8–11
Morus—Mulberry	✓		✓			4–10
Moschosma			✓			10–12
Muehlenbeckia—Muhly Grass	✓		✓			8–11
Murraya			✓			10–12
Musa—Banana	✓	✓				9–12
Muscari—Grape Hyacinth	✓	✓			✓	4–10
Mutisia—Climbing Gazania			✓			8–11
Myoporum—Boobialla, Ngaio	✓		✓			9–11
Myosotidium—Chatham Island Forget-me-Not	✓					9–11
Myosotis—Forget-me-Not	✓	✓	✓			5–10
Myrrhis—Sweet Cicely		✓				5–10
Myrsine			✓			9–11
Myrtus—Myrtle	✓		✓			8–11
Nandina—Heavenly Bamboo			✓			5–10
Narcissus—Daffodil					✓	4–9
Nelumbo—Sacred Lotus	✓	✓				6–12
Nemesia	✓		✓			8–11
Nemophila	✓					7–11
Neomarica		✓			✓	10–11
Nepeta—Catnip, Catmint	✓	✓	✓			3–10
Nephrolepis—Ladder Fern, Boston Fern		✓				10–12
Nerine—Spider Lily	✓				✓	8–10
Nerium—Oleander			✓			9–11
Nertera—Bead Plant	✓					8–11
Nestegis	✓		✓			9–12
Nicotiana—Tobacco	✓	✓				8–11/*
Nierembergia—Cup Flower	✓	✓	✓			8–10
Nigella	✓					6–10
Nomocharis	✓				✓	7–10
Nothofagus—Southern Beech	✓					7–10
Notholirion					✓	7–10
Notospartium—New Zealand Pink Broom	✓		✓			8–10
Nymphaea—Water Lily		✓				4–12
Nyssa—Tupelo	✓		✓	✓		4–10
Ochna—Bird's Eye Bush, Mickey Mouse Plant	✓					9–11
Ocimum—Basil	✓					10–12/*
Odontoglossum		✓				10–12
Oenothera—Evening Primrose	✓	✓	✓			4–10
Oldenburgia	✓					9–11
Olea—Olive			✓			8–12
Olearia—Daisy Bush	✓		✓			8–10
Omphalodes—Navelwort	✓	✓				6–9
Ophiopogon—Mondo Grass, Lilyturf	✓	✓				6–10
Opuntia—Prickly Pear	✓		✓			8–11
Origanum—Marjoram, Oregano	✓	✓	✓			7–11

Plant	Seed	Divide	Cutting	Bud/Graft	Bulb	Zone
Ornithogalum—Chincherinchee, Star-of-Bethlehem	✓				✓	6–11
Orphium—Sticky Flower	✓					9–11
Orthrosanthus	✓	✓				9–10
Osmanthus			✓			6–11
Osteospermum			✓			8–10
Ostrowskia—Giant Bellflower		✓				7–9
Ostrya—Hop Hornbeam	✓					2–9
Ourisia	✓	✓				7–9
Oxalis—Wood Sorrel, Shamrock		✓			✓	5–11
Oxydendrum—Sorrel Tree, Sourwood	✓		✓			3–9
Oxylobium	✓		✓			9–10
Pachysandra		✓	✓			5–10
Pachystachys—Golden Candles			✓			10–12
Pachystegia—Marlborough Rock Daisy	✓		✓			8–10
Paeonia—Peony	✓	✓		✓		5–9
Panax—Ginseng	✓	✓				6–9
Pancratium—Sea Lily					✓	8–11
Pandorea—Bower Vine, Wonga Wonga Vine			✓			9–11
Papaver—Poppy	✓	✓				4–10/*
Paphiopedilum—Slipper Orchid, Lady's Slipper		✓				10–12
Parahebe	✓	✓	✓			8–10
Paraserianthes—Cape Leeuwin Wattle	✓					9–11
Paratrophis	✓		✓			9–11
Parrotia—Persian Witch Hazel, Persian Ironwood	✓					5–9
Parsonsia	✓		✓			9–10
Parthenocissus—Virginia Creeper, Boston Ivy			✓			3–10
Passiflora—Passionflower	✓		✓			8–12
Pastinaca—Parsnip	✓					7–10/*
Paulownia—Princess Tree, Foxglove Tree	✓		✓			5–10
Pelargonium—Geranium	✓		✓			9–11
Pellaea—Brake Fern		✓				9–12
Pennantia			✓			9–10
Penstemon	✓	✓	✓			3–10
Pentas—Star Cluster	✓		✓			10–12
Pericallis—Cineraria	✓					9–11/*
Perovskia—Russian Sage	✓		✓			5–9
Persea—Avocado	✓		✓	✓		10–11
Persicaria—Knotweed		✓				3–10
Persoonia—Geebung	✓		✓			10–11
Petroselinum—Parsley	✓					5–11
Petunia	✓		✓			9–11/*
Phacelia—Scorpion Weed	✓					8–11/*
Phaenocoma			✓			9–11
Phaseolus—Beans	✓					8–11/*
Phebalium	✓		✓			9–11
Phellodendron—Cork Tree	✓		✓	✓		3–9
Philadelphus—Mock Orange	✓		✓			3–10
Philodendron		✓	✓			10–11

Plant	Seed	Divide	Cutting	Bud/Graft	Bulb	Zone
Phlomis	✓	✓	✓			7–10
Phlox	✓	✓	✓			4–10
Phoenix—Date Palm	✓					9–12
Phormium—New Zealand Flax	✓	✓				8–11
Photinia (incl. *Stranvaesia*)			✓			5–10
Phygelius—Cape Fuchsia	✓	✓	✓			8–11
Phylica—Flannel Flower, Cape Myrtle	✓		✓			9–11
Phyllocladus—Celery Pine	✓					8–10
Phyllodoce	✓		✓			2–9
Phyllostachys	✓	✓				5–11
Physalis—Ground Cherry	✓	✓				6–10/*
Physostegia—Obedient Plant	✓	✓	✓			3–10
Phyteuma—Rampion	✓	✓				6–10
Phytolacca—Pokeweed, Pokeberry	✓					3–11
Picea—Spruce	✓			✓		1–9
Pieris—Lily-of-the-Valley Shrub	✓		✓			4–10
Pileostegia			✓			9–10
Pimelia—Rise Flower	✓		✓			8–11
Pimpinella—Anise, Aniseed	✓					5–10/*
Pinus—Pine	✓					2–11
Pisonia	✓		✓			10–12
Pittosporum	✓		✓			8–11
Pisum—Pea	✓					*
Plagianthus—Ribbonwood	✓		✓			7–9
Platanus—Plane, Sycamore			✓			3–10
Platycodon—Balloon Flower	✓	✓	✓			4–10
Plectranthus			✓			9–12
Pleione		✓				8–10
Plumbago—Leadwort	✓		✓			9–12
Plumeria—Frangipani			✓			10–12
Podalyria—Sweet Pea Bush	✓		✓			9–10
Podocarpus—Plum Pine	✓		✓			8–12
Podolepis	✓					9–10/*
Podophyllum	✓	✓				3–9
Podranea—Port St. John Creeper			✓			9–11
Polemonium—Jacob's Ladder	✓	✓	✓			3–9
Polianthes—Tuberose		✓			✓	9–11
Polygala—Milkwort			✓			6–11
Polygonatum—Solomon's Seal		✓				3–9
Polygonum—Knotweed	✓	✓	✓			3–10
Polypodium—Polypody			✓			3–11
Polystichum		✓				3–11
Pomaderris—Tainui	✓		✓			8–10
Pontederia—Pickerel Weed		✓				3–10
Populus—Poplar, Aspen, Cottonwood			✓			1–10
Portulaca	✓					9–11/*
Portulacaria—Jade Plant, Elephant's Food			✓			9–11
Posoqueria			✓			10–12

Plant	Seed	Divide	Cutting	Bud/Graft	Bulb	Zone
Potentilla—Cinquefoil	✓	✓	✓			3–9
Pratia		✓	✓			8–10
Primula—Primrose, Polyanthus	✓	✓				4–9/*
Prostanthera—Mint Bush			✓			8–10
Protea—Sugarbush	✓		✓			9–10
Prunus—Cherry, Plum, Almond, Apricot, etc.	✓		✓	✓		2–10
Pseudopanax—Lancewood	✓		✓			9–11
Pseudowintera—Pepper Tree	✓		✓			8–10
Psidium—Guava	✓			✓		9–12
Psoralea—Blue Pea			✓			9–11
Ptelea—Hop Tree, Water Ash	✓			✓		2–9
Pteris—Brake Fern		✓				9–12
Pterocarya—Wingnut	✓		✓	✓		5–10
Pterostylis—Greenhood Orchid		✓				9–11
Pterostyrax—Epaulette Tree	✓					4–9
Pulmonaria—Lungwort	✓	✓	✓			4–9
Pulsatilla—Pasque Flower	✓	✓	✓			4–9
Punica—Pomegranate	✓		✓			9–10
Puya	✓					8–10
Pyracantha—Firethorn			✓			5–10
Pyrostegia—Brazilian Flame Vine			✓			9–11
Pyrus—Pear	✓			✓		3–10
Quercus—Oak	✓		✓	✓		3–10
Quintinia	✓		✓			9–10
Ranunculus—Buttercup	✓	✓				3–10
Raoulia—Scabweed, Vegetable Sheep		✓				7–9
Raphanus—Radish	✓					6–10/*
Ratibida—Mexican Hat	✓	✓				3–9
Rebutia—Crown Cactus, Hedgehog Cactus		✓				9–11
Rehmannia—Chinese Foxglove	✓	✓				9–10
Reinwardtia—Yellow Flax		✓	✓			9–11
Reseda—Mignonette	✓					6–10/*
Rhabdothamnus	✓		✓			10–11
Rhamnus—Buckthorn			✓			3–10
Rhaphiolepis—Indian Hawthorn			✓			8–11
Rheum—Rhubarb		✓				4–9
Rhododendron	✓	✓	✓	✓		3–11
Rhodohypoxis—Rose Grass		✓				8–10
Rhodotypos	✓		✓			5–9
Rhopalostylis—Nikau Palm	✓					10–11
Rhus—Sumac	✓	✓	✓			2–11
Ribes—Currant			✓			2–9
Ricinus—Castor Oil Plant, Castor Bean	✓					9–11
Robinia—Locust, False Acacia	✓		✓	✓		3–10
Rodgersia	✓	✓				5–9
Rohdea		✓				7–10
Romneya—California Tree Poppy	✓		✓			7–10
Romulea	✓	✓			✓	7–10

Plant	Seed	Divide	Cutting	Bud/Graft	Bulb	Zone
Rosa—Rose	✓		✓	✓		3–10
Rosmarinus—Rosemary	✓		✓			6–10
Rubia—Madder	✓	✓				6–11
Rubus—Bramble. Raspberry, Blackberry		✓	✓			2–10
Rudbeckia—Coneflower	✓	✓	✓			3–10
Rumex—Sorrel, Dock	✓	✓				3–10
Rumohra		✓				9–12
Ruscus—Butcher's Broom	✓	✓	✓			7–10
Russelia—Coral Plant			✓			9–12
Ruta—Rue	✓	✓	✓			5–10
Sagina—Pearlwort, Irish Moss	✓	✓				4–10
Salix—Willow, Osier			✓			2–10
Salpiglossis	✓					9–11/*
Salvia—Sage	✓	✓	✓			5–12
Sambucus—Elderberry	✓		✓			3–10
Sandersonia—Golden Lily-of-the Valley	✓	✓				8–11
Sanguinaria—Bloodroot		✓				3–9
Sanguisorba (incl. *Poterium*)—Burnet		✓				4–9
Sansevieria—Bowstring Hemp		✓	✓			10–12
Santolina—Lavender Cotton	✓		✓			7–10
Sapium—Tallow Tree			✓			8–11
Saponaria—Soapwort	✓	✓	✓			4–10
Sarcococca—Sweet Box	✓		✓			6–10
Sasa—Bamboo		✓				7–11
Satureja—Savory	✓	✓	✓			6–10/*
Saxifraga—Saxifrage	✓	✓				3–10
Scabiosa	✓	✓	✓			4–10
Scaevola—Fan Flower			✓			9–11
Schefflera (incl. *Dizygotheca*) Umbrella Tree	✓		✓			9–12
Schinus—Pepper Tree	✓		✓			9–12
Schizanthus—Poor Man's Orchid	✓					*
Schizostylis—Crimson Flag		✓			✓	7–10
Schotia—Boerboon	✓					9–12
Sciadopitys—Umbrella Pine	✓					5–10
Scilla		✓			✓	5–10
Scirpus		✓				6–10
Scutellaria—Skullcap, Helmet Flower	✓	✓				5–12
Sedum—Stonecrop	✓	✓	✓			6–12
Sempervivum—Houseleek	✓	✓				5–10
Senecio	✓	✓	✓			6–12
Senna	✓		✓			6–12
Sequoiadendron—Giant Sequoia, Redwood	✓		✓			7–10
Serissa	✓		✓			9–12
Serruria	✓		✓			9–11
Sesamum—Sesame	✓					9–12
Shortia		✓	✓			5–9
Sidalcea—Prairie Mallow		✓				6–10
Silene—Campion, Catchfly	✓	✓	✓			2–10

Plant	Seed	Divide	Cutting	Bud/Graft	Bulb	Zone
Sinningia—Gloxinia	✓		✓			11–12
Sisyrinchium	✓	✓				3–10
Skimmia			✓			7–10
Smilacina—False Solomon's Seal		✓				3–9
Solanum	✓	✓	✓			5–12/*
Soleirolia—Baby's Tears		✓	✓			9–11
Solenostemon (*Coleus*)	✓		✓			10–12
Solidago—Goldenrod	✓	✓				3–10
Sollya—Australian Bluebell			✓			9–11
Sophora—Kowhai, Pagoda Tree	✓		✓			5–11
Sorbus—Rowan, Whitebeam	✓		✓			2–9
Sparaxis—Harlequin Flower	✓				✓	9–10
Sparmannia—African Hemp			✓			9–11
Spartium—Broom, Spanish Broom	✓		✓			6–11
Sphaeralcea—Globe Mallow	✓	✓	✓			5–10
Spinacia—Spinach	✓					5–10
Spiraea			✓			4–10
Sprekelia—Jacobean Lily					✓	9–11
Stachys—Betony	✓	✓				5–10
Stachyurus			✓			5–10
Staphylea—Bladdernut			✓			5–9
Stauntonia—Staunton Vine	✓		✓			8–10
Stenocarpus—Queensland Firewheel	✓					9–12
Stephanandra			✓			4–10
Stephanotis—Madagascar Jasmine	✓		✓			10–12
Sternbergia—Autumn Crocus		✓			✓	7–10
Stewartia			✓			6–10
Stokesia—Stokes' Aster	✓	✓	✓			7–10
Strelitzia—Bird of Paradise	✓	✓				10–12
Streptocarpus—Cape Primrose	✓	✓	✓			10–11
Streptosolen—Marmalade Bush			✓			9–11
Styrax—Snowbell	✓					6–10
Symphoricarpos—Snowberry			✓			3–10
Symphytum—Comfrey		✓	✓			5–10
Syringa—Lilac		✓	✓	✓		4–9
Tagetes—Marigold	✓					*
Tamarix—Tamarisk			✓			2–10
Tanacetum—Pyrethrum, Tansy, Feverfew	✓	✓	✓			4–10
Taxodium—Swamp Cypress	✓		✓			6–10
Taxus—Yew	✓		✓			5–10
Tecoma			✓			10–12
Tecomanthe	✓		✓			9–11
Tecomaria—Cape Honeysuckle			✓			10–12
Tecophilea—Chilean Crocus					✓	8–9
Tellima—Fringe Cups		✓				6–9
Telopea—Waratah	✓		✓			9–11
Ternstroemia			✓			7–10
Tetrapanax—Rice Paper Plant	✓	✓	✓			8–11

Plant	Seed	Divide	Cutting	Bud/Graft	Bulb	Zone
Tetrapathea—New Zealand Passionflower	✓		✓			9–10
Teucrium—Germander	✓	✓	✓			6–11
Thalictrum—Meadow Rue	✓	✓				6–10
Thryptomene	✓		✓			9–11
Thuja—Arborvitae			✓			4–10
Thujopsis—Mock Thuja			✓			5–10
Thymus—Thyme	✓	✓	✓			5–10
Tiarella—Foamflower	✓	✓				3–10
Tibouchina—Lasiandra, Glory Bush			✓			10–12
Tigridia—Tiger Flower, Jockey's Cap	✓				✓	8–10
Tilia—Lime, Linden	✓		✓	✓		3–10
Tolmiea—Piggyback Plant		✓	✓			7–10
Toona (*Cedrela*)—Toon			✓			6–12
Torenia—Wishbone Flower	✓					*
Toronia—Toru	✓		✓			9–10
Townsendia	✓					3–10
Trachelium		✓				8–10
Trachelospermum			✓			8–11
Trachymene (*Didiscus*)	✓					9–12/*
Tradescantia (incl. *Rhoeo*)—Spiderwort	✓	✓	✓			7–12
Tragopogon—Salsify		✓				3–10/*
Tricyrtis—Toad Lily		✓				5–9
Trigonella—Fenugreek	✓					*
Trillium—Wakerobin, Trinity Flower	✓	✓				4–9
Tristania—Water Gum			✓			9–11
Triteleia					✓	4–9
Tritonia					✓	9–10
Trollius—Globe Flower		✓				5–9
Tropaeolum—Nasturtium, Canary Bird Flower	✓	✓				8–11
Tsuga—Hemlock	✓		✓	✓		4–10
Tulbaghia		✓			✓	7–10
Tulipa—Tulip	✓				✓	5–9
Tweedia (*Oxypetalum*)	✓		✓			9–11
Ulmus—Elm	✓		✓	✓		3–10
Umbellularia—California Laurel	✓		✓			7–10
Urceolina		✓			✓	9–10
Uvularia—Bellwort		✓				3–9
Vaccinium—Cranberry, Blueberry	✓	✓	✓			2–10
Valeriana—Valerian	✓	✓	✓			3–9
Vancouveria		✓				5–9
Veltheimia—Veldt Lily, Cape Hyacinth	✓				✓	9–11
Verbascum—Mullein	✓	✓	✓			4–10
Verbena	✓	✓	✓			4–10/*
Veronica—Speedwell	✓	✓	✓			3–10
Vestia			✓			9–11
Viburnum	✓		✓			3–10
Viguiera	✓		✓			9–11
Viminaria—Golden Spray			✓			8–10

Plant	Seed	Divide	Cutting	Bud/Graft	Bulb	Zone
Vinca—Periwinkle	✓		✓			4–10
Viola—Violet, Pansy	✓	✓	✓			2–10/*
Virgilia—Keurboom	✓		✓			9–11
Vitex—Chaste Tree	✓		✓			8–12
Vitis—Grape			✓	✓		5–10
Wachendorfia—Red Root	✓	✓				8–11
Wahlenbergia—Rock Bell		✓				7–11
Watsonia					✓	7–10
Weigela			✓			4–10
Weinmannia—Kamahi	✓		✓			9–12
Widdringtonia—African Cypress			✓			9–11
Wisteria	✓		✓			5–10
Xeranthemum—Immortelle	✓					*
Xeronema—Poor Knight's Lily	✓	✓				10–11
Yucca	✓	✓				6–12
Zantedeschia—Arum Lily, Calla Lily	✓	✓			✓	8–11
Zea—Corn, Maize	✓					7–11/*
Zelkova	✓		✓	✓		5–10
Zenobia			✓			5–10
Zephyranthes—Zephyr Flower	✓				✓	7–12
Zinnia	✓					5–11/*
Zoysia		✓				10–12
Zygopetalum		✓				10–12

4
Growing from seed

Seeds are amazing. They contain all the genetic information necessary to produce adult plants, from tiny alpines to forest giants. They come in all shapes and sizes, from the dust-like grains of begonias through to giant palm seeds, yet they are all essentially similar in structure and all are perfectly adapted to the environment in which they evolved.

Seed structure

Within the outer coat, a seed contains an embryo and an endosperm (a food storage organ). Seed coats vary considerably: some are hard, black and shiny, others are beautifully patterned and some are interestingly textured. Regardless of appearance, though, they all serve the same function of protecting the embryo inside.

The endosperm is a concentrated store of the fats, carbohydrates and/ or oils essential for keeping the developing embryo alive until it has functioning roots. It often makes up the bulk of the seed and is something we are all familiar with, as it is the part we consume when we eat peas, beans and nuts.

Seeds come in a wide range of sizes and shapes, which affects how they are sown.

47

The embryo, which is usually very small in relation to the seed, is the young plant. If the embryo is damaged the seed is useless. The other parts of the seed are only there to protect or nourish the developing embryo.

The size of a seed is not directly related to the final size of the plant. While there are few small plants with large seeds, many forest giants germinate from something considerably smaller than your thumbnail.

Obtaining seed

Other than being given them, there are three ways to obtain seeds: buy them, do your hybridizing or collect naturally occurring seeds.

Buying seed

The most convenient way to obtain seed is to buy it. Reputable seed companies will ensure that their seed is accurately named and true to type. The seed will have already been cleaned and will be ready for sowing or pre-sowing treatment. Also, seed packets often include valuable information about germination time, temperature, average germination percentage and the viability time. Buying seed may not be as satisfying as collecting your own, but it saves a lot of guesswork and time.

"Impulse" series impatiens, typical of the consistent modern F1 hybrid seedling strains of bedding plants.

F1 hybrids
You will often find packets of seeds labeled F1 (first filial) hybrids. These plants result from a cross between two species that always produce a consistent hybrid offspring in the first or F1 generation. Each new batch of F1 seed is the result of repeating that cross and there is no point in saving the seed of F1 hybrids because they will not reproduce true to type in the second or F2 generation.

Hybridizing

For a plant propagator there is nothing more satisfying than producing a new hybrid that is clearly an improvement on anything previously available. It is not a quick process, though, and many hybridizers labor for years without producing anything outstanding.

Hybridization is just a matter of transferring pollen from one parent to the intended seed-bearing parent, but some plants have more accessible flower parts than others. Normally, only closely related plants will successfully hybridize, usually those within the same genus, although there are a few examples of intergeneric hybridizing. Practice with something simple, such as fuchsias or lilies.

One of the most common hybridizing problems occurs with double flowers. Sometimes the stamens of double flowers become modified into

A tulip showing flower structure.

48

petals, thereby rendering the flower sterile for pollen production. Such flowers are generally suitable as seed parents only, although they may occasionally produce a functioning anther.

Problems also arise when crossing plants that flower at different times. The only solution is to preserve some pollen until the seed parent is in flower. For this use some gelatin medicine capsules, a desiccant (silica gel is the most widely available), a shallow jar or some other small container. When the pollen parent is in flower, remove a few fresh stamens and trim them down to just the anthers and a short length of filament so that they will fit in the medicine capsules. Put a thin layer of desiccant in the jar and place the capsules on the desiccant. Keep the jar, unsealed, in the refrigerator. The gelatin capsules are semi-permeable and the moisture from the stamens will be drawn out by the desiccant, which you should replace daily. After a few days the stamens will have become thoroughly dry. Once dried, refresh the desiccant and seal the container. The dried pollen can then be stored, usually indefinitely, in the freezer until needed. Allow the pollen to warm before use.

To become a serious hybridizer you must also be aware of the many interwoven genetic factors that affect your crosses. These range from quite mundane things, such as crosses that fail because of incorrect pollen size, through variations in chromosome count, to such arcane things as thrum homozygote mortality. A thrum, incidentally, is a form of *Primula* flower in which the stigma is on a very short style.

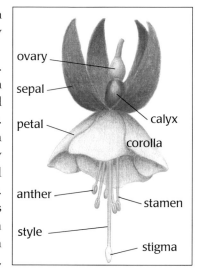

Fuchsias also have a very obvious flower structure.

Collecting seed

The cheapest way to obtain seed is to collect it from garden plants or those growing in the wild, and while you could simply harvest whatever ripe seed is available, it is usually better to plan ahead. When your chosen plants have finished flowering look for any swelling seed pods or ripening fruit. While fertile fruits are usually quite apparent, determining just when the seed is ripe is a matter of experience and judgment. If there is any doubt, many seeds can be collected slightly unripe and will continue to ripen off the plant provided they remain in their seedpods.

To extract the seed from dry pods and cones, hang them upside down over a paper bag in a dry airy place and wait for the seed to fall. An occasional gentle tap will loosen them. Alternatively, keep the seedpods or cones in a saucer and break them up as they start to dry, thereby releasing the seed.

Extracting the seed from soft fruits can be awkward and messy. The easiest way is usually to crush the fruit, then steep it in water to remove the pulp. Tip off or lift out any floating material and the reasonably clean seed should be left at the bottom of the container. The fruit may need to be left in water for several days (this also softens the seed coat), and even

Hybridization:
[a] Lily anther with a heavy coating of pollen.
[b] Lily pollen being transferred from the anther to the sticky-coated stigma.
[c] Very immature seeds developing in a lily seedpod.

The seed-filled fruit of the pomegranate (*Punica granatum*).

then the seed may not come away completely cleanly.

Whenever you harvest your own seed, it is important to remove as much of the chaff and surplus vegetable material as possible. If this is sown along with the seed it tends to rot and may encourage fungal diseases.

Storing seed

Warmth and moisture are the greatest enemies of stored seed. If you cannot sow seed immediately, it should be kept somewhere cool and dry. Dust the seed with a fungicide and keep it in small paper bags in the bottom of your refrigerator. Paper bags, unlike plastic, allow the moisture to escape from the seed and the cool temperature of the refrigerator slows down the natural deterioration. Very fleshy seeds that become desiccated under such conditions are best kept in plastic bags.

Germination and viability

Many seeds require precise conditions for germination and the viability period of seed varies greatly. Seeds of plants that grow where rainfall is sporadic or virtually non-existent usually have an extremely long viability period, as the seed has to survive in a dormant state until the next rain. Likewise the seed of plants from areas with long, cold winters has to be able to survive the winter. Some seed actually needs to be exposed to the cold of winter before it will germinate.

Cones of ponderosa pine or western yellow pine (*Pinus ponderosa*).

50

How a plant disperses its seeds may also have an influence on their germination and viability. For example, many seeds are enclosed in soft-bodied fruit, and may not germinate until their hard seed coating has been softened. In the wild this usually happens when the seed passes through the digestive system of an animal.

Breaking dormancy

Many seeds are difficult to germinate unless you know how to break their dormancy. Commercial seed often has information on the seed packet, but when in doubt, don't be afraid to experiment. Usually the plant will provide you with clues. For example, fast-growing annuals will normally germinate rapidly, requiring nothing more than the right temperature and an initial watering to start them off, but plants that come from extreme climates may have particular requirements. Those from very cold areas may require stratification; those subject to prolonged drought may need soaking, and very hard-coated seeds may need scarification as well as soaking.

Hellebore seedlings. Hellebore seeds often require two periods of stratification.

Stratification

Stratification, which is exposing the seed to cold, simulates natural winter chilling. Moisten some starter (seed-raising) mix, blend in the seeds and then put the seeds and soil in a plastic bag, or better still, a translucent plastic box with a lid. Leave the bag somewhere fairly warm, out of direct sun, for three or four days to allow the seeds to take up moisture and then transfer it to the refrigerator (not the freezer). Just how long the seed will need to be chilled varies. ten weeks is about the average. Give the container a shake occasionally to make sure the soil remains loose and aerated. If the seeds have started to germinate, it is definitely time to get them out of the refrigerator. If, after stratifying, the seed does not germinate, yet still appears quite sound, try a repeat period of chilling—some seeds have to be stratified twice.

Scarification

Scarification is just abrading the seed coat in order to allow it to soften more quickly when moistened. To scarify small seeds, line a jar with coarse sandpaper, put the seed inside, screw on the lid, then shake them until the seed coat is well scratched. Larger or more stubborn seed coats can be rubbed with a file or nicked with a sharp blade, but be careful not to damage the embryo.

Soaking

Some seeds have a very hard seed coat that responds better to softening in hot water rather than by abrading. Other seeds, especially those of plants from areas with very irregular rainfall contain germination inhibitors that must be leached out. If these seeds were to germinate after a light shower they would certainly perish, so they will not germinate until they have had a really thorough soaking.

To soften seeds use moderately warm, not boiling, water. Twelve hours' soaking at a temperature of around 113°F (45°C) is usually adequate, after which they may have swelled slightly.

A few seeds with particularly tough seed coats require soaking in a mild acid solution to soften them. This simulates the effect of passing through an animal's digestive tract. Use very dilute hydrochloric or phosphoric acid and exercise extreme caution. This is definitely a last resort.

In recent years it has become clear that seeds normally released in bush fires often need to be exposed to smoke to initiate chemical changes that result in the high germination rates seen in areas regenerating after fire. Because smoking the seed is not always easy or practical and it is seldom possible to obtain the right sort of burning material from the seeds' natural range, some companies now offer powdered mixes that can be added to the soaking water or starter mix. These mixes contain either powdered or granular water-soluble ash and/or chemical extracts from ash that will help break dormancy.

Germination temperature

All seeds have an optimum temperature range at which they germinate best, as well as thresholds, above and below which they will not germinate at all. Fortunately, most seeds germinate well over a wide range of temperatures.

If you intend to germinate seeds in spring and summer only, artificial heating may be unnecessary. However, if you wish to give your seeds an early start or keep them growing on well into winter, heating pads will be essential. When using any sort of soil heating, make regular moisture checks as the additional heat can quickly dry the soil.

Starter mixes

For just a few seeds it is probably more convenient to use a starter mix straight from the bag, but most commercial seed-raising mixes are less than ideal. You will get far better results and a more consistent mix if you make your own.

By all means, use one of the bagged mixes as your base, but find one that is bark-, peat- or fern fiber-based, reasonably consistent and well composted. You do not want a mix that is too fresh, poorly composted or showing signs of fungus or molds. Once you have a good basic mix you can adjust it to your own specifications.

The starter mix should be light and airy, yet moisture retentive. I add finely chopped sphagnum moss because it contains natural fungicides, retains huge amounts of moisture while remaining open and free from compaction, and is easily penetrated by the fine roots of young seedlings. To further improve the aeration and drainage, I use perlite or fine pumice.

The final mix should be about one-third each of bark- or peat-based potting mix, fine sphagnum moss and perlite or pumice. The sphagnum is quite fibrous, so the mix must be sieved. Otherwise, the roots of the young seedlings will get caught up in the strands of moss, which makes them difficult to transplant. Run the mix through a ¼ in (6 mm) mesh soil sieve. Do not wet it first or it will be next to impossible to sieve.

You can use substitutes for these ingredients and as you gain experience you will almost certainly develop your own mixes. Avoid garden soil because if not sterilized it will be full of weeds and diseases. Also, it will compact down like concrete. A little sand will open up the mix and improve the drainage, but too much and the mix becomes hard and caked. Some propagators use vermiculite, but I find it rapidly breaks down to a soggy mass, encourages algae growth and makes transplanting difficult.

Sowing carrot seed, which is sown very near the surface rather than in furrows.

Seed sowing

The type of seed and its intended purpose largely determines whether you sow your seed in containers or in the open ground. Most vegetable seed and many flowering annuals can be sown directly where they are intended to grow. These plants germinate and grow so reliably that they can just be left to get on with it.

Container growing under cover has the advantage of a more easily controlled environment, but the seedlings will have to be transplanted at least once (every transplant is a shock and will check growth to at least a small degree) and they will require regular watering and occasional feeding. It is better to sow in containers if you want to sow seed out of season or are unsure of how the seed will germinate and develop, or if the plants will be too tender to survive outdoors.

Sowing parsnip seeds in furrows.

Sowing outside

Preparation

When sowing in the open ground it is vitally important to have a well-prepared seedbed and sensible crop layout.

The seedbed does not need to be very deep but in order to ensure good root development there should be at least 6 in (15 cm) of loose topsoil. Prepare it by digging over the soil, adding peat, compost or other soil conditioners, then raking or hoeing the bed to fine tilth.

Fertilizer is most easily applied before the plants appear. Add a light dressing of general garden fertilizer and rake it into the soil. If you have an acid soil, add a little lime. Dolomite lime (which has extra magnesium and other trace elements) is particularly good, although hydrated lime will do. While most vegetables, particularly brassicas, prefer a slightly alkaline soil, avoid over-liming. A general dressing every two years is perfectly adequate and any plants that need lime can always be given extra. If you have any doubts about the nature of your soil, have it tested.

If you are planting out a vegetable garden, make sure that tall crops, such as sweet corn, will not shade out lower growers and that rampant growers, such as pumpkins, have adequate room to spread.

Sowing

Use a string line to mark out the rows, then with the edge of a hoe open a shallow trench approximately 2 in (5 cm) deep, depending on seed size. Sow the seed by shaking it carefully from the packet or by using a small hand sower (available from most garden suppliers). Aim for a close, even spread of seed along the row. Larger seeds like runner beans are usually individually placed in position by hand. Very fine seed is normally sown quite heavily and thinned after germination. The thinnings can often be transplanted to make new rows. Once sown, gently close the earth back over the seed to cover it and then gently water the seedbed. Keep it moist until there are signs of germination. Once the young plants are up, watch out for the signs of fungus troubles, slugs and snails.

Occasionally, you may find that the soil becomes caked and the seedlings have trouble pushing up through the surface crust. If this is a regular problem with your soil, try mixing in sand with the soil you use to cover the seed. A couple of handfuls of coarse river sand every yard or so should prevent crusting.

A closely planted and very productive small home vegetable garden.

Sowing in containers

Propagating trays are the best containers for raising seeds, although almost any shallow box or tray that will hold soil and has drainage holes will do. Fill the container to just less than 1 in (2.5 cm) from the top (this space allows the seedlings room to develop), then gently firm the mix into place and water with a fine mist. The tray is now ready for sowing.

Getting the seed evenly distributed over the tray is often the trickiest part of sowing. Very fine seed causes the most problems because it is hard to see. If you mix the seed with plain talcum powder or icing sugar, the whiteness of these substances against the dark soil will highlight the distribution pattern.

To cover or not to cover?

Beginners tend to bury their seeds beyond recovery. In most cases all that is required is a very fine dusting of soil, if any at all. Some seeds need light to germinate and must not be covered. A good rule is to cover the seed to just over its own depth, so fine seed is only lightly covered while larger seeds are buried more deeply. Very fine seed, such as that of begonias and most ericaceous plants, should not be covered at all. Just keep it moist and it will germinate on the surface and send down roots.

Cover the seed by lightly and evenly dusting it with finely sieved starter mix. Water it with a very fine mist or by soaking from below, then cover the tray with a pane of glass and place it in a warm place out of direct sun. Cover the glass with a sheet of newspaper to shade the seed; even seeds that need light for germination should be shaded or the heat of the sun through the glass may cook them.

Aftercare

Seedlings are at their most vulnerable immediately after they have germinated when they are very easily destroyed by slugs, snails and birds, and may collapse with fungal diseases if too wet. Uncover the seedlings by gradually raising the glass cover (prop it up with a stone, twig or piece of wire) and then finally removing it. Good ventilation, combined with a mild fungicide, should control most damping-off diseases.

Once your seedlings have their first true leaves it is time to transplant them from the seed tray into less crowded containers. If the young plants are very small, transplant them into a finely sieved 50/50 blend of potting mix and perlite, otherwise an ordinary all-purpose potting mix is fine.

Transplanting can be delayed but you must feed the seedlings. Starter mixes do not contain much fertilizer and rapidly growing young seedlings will quickly expend their natural reserves. A mild all-purpose liquid fertilizer will keep them happy for a while, but transplant them out as early as possible.

[a] Seedlings that have been pre-soaked on moist paper towels before sowing.
[b] Pre-sprouted seedlings after being planted out. Keep them in a shaded but well-lighted place to encourage growth while preventing sunburn.

Remove any weeds as they appear, water the seedlings and feed them well. Most importantly, if they are naturally quick-growing plants, avoid allowing them to become pot-bound.

Pre-sprouting

Very large seeds, particularly monocotyledons like palms and conifers, often send down a long taproot (radicle) that either goes through the drainage holes in the seed tray or is damaged by transplanting. Such seeds are easier to handle if they are germinated before they are planted out. Soak the seeds for a few hours, place them on a tray lined with wet paper towels, cover them with more wet paper towels and put the tray on a heating pad or in a warm dark place such as the cupboard near a hot water heater. Keep the paper towels damp and within a few days the radicle should appear and the sprouted seeds can be planted into individual pots of regular potting mix.

Propagating from spores

Spores are the way that many primitive plants, such as ferns, fungi and mosses, propagate themselves. Although seemingly seed-like, they are really quite different. Ferns, the most widely grown spore-bearing plants, can be propagated by vegetative means, but if large quantities are required and time is no object, spores may be used.

Sori (spore capsules) on the underside of a fern leaf.

Collecting spores

Many ferns produce separate fertile and sterile fronds. The fertile fronds are very distinctive as their undersides become covered in blackish brown sporangia or sori (spore capsules). As the sporangia mature they rupture, releasing the spores, which can be collected by tapping a ripe fertile frond while holding a paper bag underneath it. Alternatively, pick a few ripe fertile fronds, put them in the bag and shake it.

Sowing spores

Fern spores do not remain viable for long and should be sown without delay.

Keeping the growing medium moist is essential when raising spores. When the spores first start to grow, they produce a small green scale called a prothallus, which contains both male and female genetic material. As the male gametes must actively swim or be carried in moisture to the female organ, the environment must be very moist.

The spores can be sown on a very finely sieved mix of sphagnum moss, peat and river sand, about one-third of each by volume. Fill a pot with very wet mix, dust the spores over the surface, gently water them, and then cover the pot with a pane of glass or plastic film. Place the pot in its saucer and fill the saucer with water. Keeping the saucer topped up with

A moss-covered boulder with many fern sporlings and seedlings. Moss is an excellent germination medium. Note, too, how the ferns are growing on an inert medium.

water ensures that the soil stays moist. Place the pot and saucer in a shady but reasonably warm position. The spores need light, but avoid exposing them to direct sun.

Brick method

Potting mix nearly always contains spores of algae, liverworts and mosses, which develop faster than the ferns. Therefore, I prefer a method that requires no potting mix and is clean and easy to use.

Find an old porous, unglazed, red clay brick. Scrub it clean using water and a very dilute bleach solution. Once clean, rinse it thoroughly to remove all traces of the bleach. This should ensure a sterile surface. Next, find a saucer or bowl large enough for the brick to sit in. Fill the saucer with water and place the brick in it, wide side down. After a few hours the top of the brick should become wet as moisture is drawn over it by capillary action. If the top does not become moist, you need a more porous brick or a deeper saucer.

Once you can keep the brick wet, sow the spores on top of it. Cover the brick with a large plastic bag or some other clear cover. Keep the saucer and brick in a cool place and make sure the saucer is constantly topped with water.

Within a few weeks a green film of prothalli will develop on the surface. Add a little liquid fertilizer at this stage and, if the moisture level is right, small ferns will soon develop. When they are large enough to handle, gently prize them off the brick and pot them.

Pasque flowers (*Pulsatilla vulgaris*) going to seed.

Table 2: Germination requirements

The following table lists all the genera indicated as being suitable for raising from seed in Table 1. The temperatures indicated are the optimum for quick and even germination, and the germination times apply only to seeds sown at temperatures within the optimum range.

Pre-sowing treatment is optional in some cases and critical in others. It is wise to carry out the pre-sowing treatment even if it's only suggested.

The covering requirements have been abbreviated to three main types:
- "Yes" if the seed must be covered and kept in a dark place until germination;
- "Lightly" if just a thin covering of soil over the seed and light shading of the tray is required;
- "No" if the seed needs light to germinate or is very fine.

Plant	Pre–sowing	Temp °F / °C	Cover	Germination
Abelmoschus	Soak 12 hours	68–77 / 20–25	Yes	10–15
Abies	Stratify 8–12 weeks	59–68 / 15–20	Yes	35–70
Acacia	Scarify or soak 24 hours	59–77 / 15–25	Yes	10–35
Acanthus		64–72 / 18–22	Yes	15–25
Acca	Remove all fruit pulp	64–75 / 18–24	Lightly	10–28
Acer	Stratify 8–12 weeks	59–72 / 15–22	Yes	14–42
Achillea		64–70 / 18–21	No	10–15
Ackama		59–72 / 15–22	Lightly	21–42
Acmena		64–75 / 18–24	Lightly	14–35
Aconitum	Stratify 3 weeks	55–64 / 13–18	Lightly	28–35
Actaea	Stratify 8–10 weeks	59–68 / 15–20	Yes	14–35
Adenandra		68–77 / 20–25	Lightly	7–21
Aesculus	Scarify & stratify 12 weeks	59–72 / 15–22	Yes	14–42
Aethionema		55–64 / 13–18	Lightly	14–21
Agapanthus		70–77 / 21–25	Yes	21–35
Agastache		59–72 / 15–22	Lightly	7–21
Agathis		64–75 / 18–24	Yes	28–60
Agave	Soak 8 hours	64–75 / 18–24	Yes	18–35
Ageratina		64–72 / 18–22	Lightly	14–28
Ageratum		73–82 / 23–28	No	7–10
Agonis		64–75 / 18–24	Lightly	14–35
Ailanthus	Better stratified 6 weeks	64–77 / 18–25	Yes	21–42
Akebia	Better stratified 4 weeks	59–72 / 15–22	Yes	21–50+
Albizia	Soak 24 hrs or scarify	70–77 / 21–25	Yes	15–30
Alcea		68–72 / 20–22	Lightly	14–21
Alchemilla		55–68 / 13–20	Lightly	10–28
Alectryon	Better soaked 8 hours	64–75 / 18–24	Yes	21–42
Allium – Chives		64–72 / 18–22	Yes	10–15
Allium – Onions and Leeks		64–72 / 18–22	Lightly	10–14
Allium – Ornamental	Stratify 4 weeks	64–70 / 18–21	Lightly	14–21

Plant	Pre–sowing	Temp °F / °C	Cover	Germination
Alnus	Stratify 12 weeks	59–72 / 15–22	Lightly	18–35
Aloe		68–82 / 20–28	Lightly	25–30
Alonsoa		61–75 / 16–24	Lightly	7–21
Aloysia (Lippia)		64–77 / 18–25	Lightly	14–21
Aopecurus		64–77 / 18–25	Lightly	10–21
Alstroemeria		55–68 / 13–20	Yes	15–50
Alyssum		59–70 / 15–21	No	7–14
Amaranthus		68–77 / 20–25	Lightly	8–12
Amaryllis		64–77 / 18–25	Lightly	14–35
Amelanchier	Stratify 12 weeks	59–72 / 15–22	Yes	14–35
Ammi		68–77/ 20–25	Lightly	7–14
Anagallis		64–75 / 18–24	Lightly	7–21
Anaphalis		64–68 / 18–20	Lightly	5–10
Anchusa		64–72 / 18–22	No	7–10
Andromeda		59–70 / 15–21	No	14–28
Androsace		59–68 / 15–20	Lightly	14–21
Anemone		59–70 / 15–21	Lightly	14–21
Anemonella		59–68 / 15–20	Lightly	14–35
Anethum		59–64 / 15–18	No	5–15
Angelica		59–70 / 15–21	Yes	14–21
Angophora		64–77 / 18–25	Lightly	14–35
Anigozanthus		64–79 / 18–26	Lightly	14–21
Annona		64–75 / 18–24	Yes	18–35
Anthemis		68–73 / 20–23	Lightly	5–10
Anthericum		68–77/ 20–25	Yes	14–28
Anthriscus		59–70 / 15–21	Lightly	7–14
Antigonon		70–77 / 21–25	Yes	14–21
Antirrhinum		64–73 / 18–23	No	7–14
Apium		64–72 / 18–22	Yes	10–20
Aptenia		64–77 / 18–25	Lightly	14–28
Aquilegia	Stratify 3 weeks	64–75 / 18–24	No	21–28
Arabis		64–73 / 18–23	No	7–15
Araucaria		64–77 / 18–25	Yes	14–42+
Araujia	Soak 8 hours	64–75 / 18–24	Yes	10–28
Arbutus	Remove all fruit pulp	59–72 / 15–22	No	14–35
Archeria		59–72 / 15–22	Lightly	21–35
Arctotheca		59–70 / 15–21	Lightly	7–28
Arctotis		59–70 / 15–21	Lightly	21–35
Arenaria		55–64 / 13–18	No	14–21
Arisaema	Perhaps stratify 8 weeks	59–72 / 15–22	Yes	14–35
Arisarum		59–68 / 15–20	Yes	14–21
Aristolochia	Soak 24 hours	64–72 / 18–22	No	28–90
Armeria	Soak 8 hours	64–70 / 18–21	Lightly	7–14
Arnica		55–64 / 13–18	Lightly	21–35
Artemisia – Russian Tarragon		59–70 / 15–21	Lightly	18–28
Artemisia – Wormwood		59–70 / 15–21	Yes	7–14
Arthropodium		59–72 / 15–22	Lightly	10–28
Arum	Better stratified 6 weeks	59–72 / 15–22	Yes	10–28

Plant	Pre–sowing	Temp °F / °C	Cover	Germination
Aruncus		55–72 / 13–22	Lightly	14–35
Asarina		70–77 / 21–25	Yes	10–15
Asclepias	Better soaked 4 hours	68–77/ 20–25	Lightly	10–28
Asparagus	Soak 8 hours	75–82 / 24–28	Yes	21–42
Asphodeline		68–77/ 20–25	Lightly	28–35
Astelia		57–68 / 14–20	Yes	14–35+
Aster		64–70 / 18–21	No	14–21
Astilbe		59–70 / 15–21	No	14–21
Astrantia		59–72 / 15–22	Lightly	10–28
Aubrieta		64–70 / 18–21	No	14–21
Aulax	Gently scarify or soak	64–77 / 18–25	Yes	14–35+
Azara		59–72 / 15–22	Lightly	14–35
Babiana		59–75 / 15–24	Lightly	10–28
Baeckia		64–75 / 18–24	Lightly	10–35
Banksia	Gently scarify or soak	64–75 / 18–24	Yes	14–35+
Baptisia	Scarify	68–77/ 20–25	Yes	5–15
Bassia		68–77/ 20–25	No	10–15
Beaufortia		64–75 / 18–24	Lightly	14–28
Beaumontia		64–77 / 18–25	Yes	14–28
Begonia		72–79 / 22–26	No	14–21
Beilschmiedia		59–75 / 15–24	Lightly	21–42
Belamcanda		68–82 / 20–28	Yes	14–21
Bellis		64–75 / 18–24	No	7–15
Berberidopsis		59–72 / 15–22	Lightly	14–35
Berberis	Stratify 8–12 weeks	59–72 / 15–22	Lightly	14–35
Bergenia		64–75 / 18–24	No	5–10
Berzelia		59–75 / 15–24	Lightly	10–28
Beta	Soak 24 hours	68–75 / 20–24	Lightly	15–20.
Betula	Stratify if stored	55–72 / 13–22	Lightly	10–28
Blandfordia		59–75 / 15–24	Lightly	15–30
Boltonia		59–72 / 15–22	Lightly	10–21
Borago		64–70 / 18–21	Yes	5–15
Bougainvillea		68–77/ 20–25	Lightly	28–50
Brachychiton		68–81 / 20–27	Lightly	18–35
Brachycome		64–70 / 18–21	No	7–10
Brassica crops		64–72 / 18–22	Yes	7–14
Briza		64–72 / 18–22	Lightly	5–10
Brodiaea		59–75 / 15–24	Lightly	14–28
Browallia		70–77 / 21–25	No	7–15
Brunnera		68–82 / 20–28	Lightly	14–21
Buddleja		68–77/ 20–25	Yes	15–30
Bulbinella		59–75 / 15–24	Yes	14–35
Bupthalmum		59–72 / 15–22	Lightly	10–25
Butia	Scarify & soak 48 hours	68–81 / 20–27	Yes	10–50+
Butomus		59–72 / 15–22	Lightly	15–25
Caesalpinia	Soak 48 hours	68–77/ 20–25	Yes	7–20
Calamintha		57–72 / 14–22	Lightly	10–21
Calandrinia	Stratify 8 weeks	57–72 / 14–22	Lightly	10–28

Plant	Pre–sowing	Temp °F / °C	Cover	Germination
Calceolaria		64–77 / 18–25	No	10–15
Calendula		64–73 / 18–23	Yes	10–15
Callicarpa	Stratify 8 weeks	59–72 / 15–22	Yes	14–35
Callistemon		64–75 / 18–24	Lightly	14–28
Callistephus		64–73 / 18–23	Lightly	8–11
Callitris		64–75 / 18–24	Lightly	14–35
Calluna	Better stratified 8 weeks	59–70 / 15–21	No	15–40
Calocedrus	Stratify 12 weeks	57–72 / 14–22	Lightly	14–35
Calochortus		59–72 / 15–22	Yes	14–35
Calodendrum		64–77 / 18–25	Yes	20–50
Calycanthus	Better stratified 8 weeks	59–72 / 15–22	Yes	14–35
Calystegia		59–72 / 15–22	Lightly	14–35
Camassia		54–68 / 12–20	Lightly	14–28
Camellia	Soak 24 hours	64–77 / 18–25	Lightly	30–70
Campanula		64–72 / 18–22	Lightly	14–28
Campsis	Stratify 10 weeks	64–73 / 18–23	Yes	14–21
Canna	Scarify & soak 24 hours	68–77 / 20–25	Yes	7–14
Capsicum		68–75 / 20–24	Yes	7–14
Caragana	Soak 24 hours or scarify	64–75 / 18–24	Yes	10–35
Cardamine	Perhaps stratify 4 weeks	54–64 / 12–18	Lightly	8–15
Cardiocrinum	Seed must be fresh	59–72 / 15–22	Yes	14–35
Carica	Remove all fruit pulp	68–77/ 20–25	Lightly	10–28
Carissa	Remove all fruit pulp	64–77 / 18–25	Yes	14–42+
Carmichaelia	Scarify or soak 24 hours	59–72 / 15–22	Yes	21–100+
Carpinus	Stratify 12 weeks	59–72 / 15–22	Lightly	21–100+
Carpodetus		59–72 / 15–22	Lightly	10–35
Carthamnus		64–72 / 18–22	No	5–15
Carum		59–77 / 15–25	Lightly	7–15
Carya	Stratify 8–12 weeks	59–72 / 15–22	Yes	14–42
Casimiroa	Remove attached flesh	64–75 / 18–24	Yes	21–35
Cassia	Scarify	68–82 / 20–28	Yes	5–14
Cassinia		59–68 / 15–20	Lightly	14–28
Castanea	Scarify & soak 24 hours	59–72 / 15–22	Yes	14–35+
Castanospermum	Soak 24 hours	68–82 / 20–28	Yes	21–42
Casuarina		68–86 / 20–30	Lightly	14–42
Catalpa	Better stratified 6 weeks	59–77 / 15–25	Lightly	14–28
Catananche		64–73 / 18–23	Lightly	5–15
Cedronella		59–72 / 15–22	Lightly	7–21
Cedrus	Better stratified 8 weeks	59–72 / 15–22	Yes	14–42+
Celmisia		55–68 / 13–20	Lightly	14–21
Celosia		72–77 / 22–25	Yes	7–10
Centaurea		64–70 / 18–21	Lightly	7–14
Cerastium		61–70 / 16–21	No	7–14
Ceratonia	Scarify & soak 24 hours	64–77 / 18–25	Yes	10–28
Ceratopetalum		64–75 / 18–24	Lightly	10–28
Ceratostigma	Better stratified 4 weeks	59–72 / 15–22	Lightly	14–35
Cercis	Soak & stratify 12 weeks	68–77/ 20–25	Yes	20–90
Cestrum	Soak at least 4 hours	64–75 / 18–24	Yes	7–21

Plant	Pre-sowing	Temp °F / °C	Cover	Germination
Chaenomeles	Stratify 12 weeks	59–72 / 15–22	Yes	14–35
Chamaedorea	Seed must be fresh	68–82 / 20–28	Yes	30–90
Chamaemelum		68–77/ 20–25	Lightly	7–14
Chimonanthus	Stratify 12 weeks	59–72 / 15–22	Yes	14–35+
Chionanthus	Seed must be fresh	59–72 / 15–22	Lightly	14–35
Chionodoxa		59–64 / 15–18	Lightly	21–35
Chordospartium	Scarify or soak 24 hours	59–72 / 15–22	Yes	14–42
Chorizema	Scarify or soak 24 hours	64–75 / 18–24	Yes	7–21
Chrysanthemum – annual		59–68 / 15–20	Lightly	10–14
Chrysanthemum – perennial		64–72 / 18–22	Lightly	7–28
Chrysanthemum maximum		64–72 / 18–22	No	10–14
Chrysanthemum parthenium		64–72 / 18–22	No	10–20
Cichorium		64–72 / 18–22	Yes	7–14
Citrullus		68–75 / 20–24	Yes	7–14
Citrus	Better if seed peeled	64–75 / 18–24	Yes	14–42
Cladanthus		68–77/ 20–25	Lightly	28–42
Clarkia/Godetia		61–72 / 16–22	Lightly	5–15
Claytonia	Stratify 4 weeks	64–72 / 18–22	Lightly	28–50+
Clematis	Stratify 4 weeks	72–82 / 22 28	Yes	30–100
Cleome		68–77/ 20–25	Lightly	10–15
Clerodendrum		64–77 / 18–25	Yes	10–28
Clianthus	Better soaked 8 hours	59–75 / 15–24	Yes	7–28
Clivia	Seed must be fresh	72–79 / 22–26	Lightly	30–80
Cobaea		59–68 / 15–20	Yes	7–10
Colchicum		59–68 / 15–20	Lightly	15–28
Consolida	Seed must be fresh	55–60 / 13–20	Yes	10–21
Coprosma		59–72 / 15–22	Lightly	10–28
Corallospartium	Scarify or soak 24 hours	59–75 / 15–24	Yes	10–28
Cordyline		59–75 / 15–24	Yes	14–35
Coreopsis – annual		68–77/ 20–25	Lightly	5–10
Coreopsis – perennial		61–72 / 16–22	No	10–15
Coriandrum		59–64 / 15–18	Lightly	7–14
Cornus	Stratify 54–68 / 12–20 weeks	64–77 / 18–25	Yes	30–300+
Corokia		55–72 / 13–22	Lightly	14–28
Coronilla	Scarify or soak 8 hours	68–77/ 20–25	Yes	15–40
Cortaderia		64–72 / 18–22	Lightly	5–10
Corylopsis	Stratify 8 weeks	59–68 / 15–20	Lightly	14–35
Corylus	Stratify 8 weeks	59–72 / 15–22	Yes	14–35
Corynocarpus		68–77/ 20–25	Yes	14–21
Cosmos		64–75 / 18–24	Yes	5–10
Cotoneaster	Stratify 8–12 weeks	59–72 / 15–22	Yes	14–35
Cotula		59–75 / 15–24	No	7–15
Crataegus	Stratify 8 weeks	59–72 / 15–22	Lightly	14–35+
Crinum		59–72 / 15–22	Yes	14–35
Cryptomeria	Better stratified 8 weeks	59–72 / 15–22	Lightly	14–35
Cucumis – Cucumber		68–75 / 20–24	Yes	5–10
Cucumis – Melon		72–79 / 22–26	Yes	7–14
Cucurbita		68–77/ 20–25	Yes	5–10

Plant	Pre–sowing	Temp °F / °C	Cover	Germination
Cuphea		64–72 / 18–22	Lightly	10–15
Cupressus	Stratify 4 weeks	64–72 / 18–22	Yes	28–42
Cyananthus		59–72 / 15–22	Lightly	14–21
Cyathodes		55–72 / 13–22	Lightly	14–42
Cyclamen – garden species		57–64 / 14–18	Yes	21–28
Cyclamen persicum – pots		57–64 / 14–18	Yes	28–35
Cydonia	Stratify 8–12 weeks	59–72 / 15–22	Yes	14–22
Cymbalaria		64–77 / 18–25	Lightly	14–28
Cymbopogon		68–77 / 20–25	No	5–15
Cynara		68–75 / 20–24	Lightly	15–25
Cynoglossum		64–70 / 18–21	Yes	5–10
Cypella		64–75 / 18–24	Lightly	10–21
Cyperus		64–77 / 18–25	Lightly	21–35
Cyphomandra		68–77 / 20–25	Lightly	14–28
Cyrtanthus	Seed must be fresh	64–75 / 18–24	Yes	10–28
Cytisus	Soak 24 hours	64–77 / 18–25	Yes	21–35
Daboecia		59–72 / 15–22	No	15–35
Dacrydium	Better stratified 8 weeks	59–72 / 15–22	Lightly	21–60+
Dahlia		59–70 / 15–21	Yes	5–10
Dais		68–77/ 20–25	Lightly	15–28
Daucus		64–72 / 18–22	Lightly	10–15
Daucus – Carrot		61–72 / 16–22	Lightly	14–21
Davidia	Stratify 12–24 weeks	59–72 / 15–22	Yes	21–42+
Decaisnea	Soak 8 hours	64–75 / 18–24	Yes	14–28
Delphinium	Seed must be fresh	59–75 / 15–24	Yes	12–18
Desfontainea		59–68 / 15–20	Lightly	21–42
Dianthus		64–72 / 18–22	Lightly	5–14
Diascia		59–72 / 15–22	Lightly	7–14
Dicentra	Stratify 6 weeks	55–64 / 13–18	Yes	21–42
Dichondra		59–77 / 15–25	Lightly	7–14
Dictamnus	Stratify 6 weeks	55–64 / 13–18	Lightly	28–42
Dierama		59–72 / 15–22	Yes	14–35
Digitalis		59–77 / 15–25	No	5–15
Dimorphotheca		59–70 / 15–21	Lightly	10–15
Diospyros	Better stratified 8 weeks	64–75 / 18–24	Yes	14–35
Dipsacus		64–72 / 18–22	Lightly	14–28
Dodecatheon	Stratify 4 weeks	59–72 / 15–22	Lightly	28–42
Dodonea		59–77 / 15–25	Lightly	145–28
Dombeya		68–77/ 20–25	Yes	14–21
Doronicum		64–72 / 18–22	No	14–21
Dorotheanthus		63–77 / 17–25	Lightly	14–21
Draba		55–64 / 13–18	No	10–28
Dracophyllum		55–72 / 13–22	Lightly	21–42
Dryas		59–72 / 15–22	Lightly	42–70
Dysoxylum		64–75 / 18–24	Lightly	14–28
Eccremocarpus		59–77 / 15–25	Lightly	7–21
Echeveria		68–81 / 20–27	No	14–35
Echinacea		68–77/ 20–25	Lightly	5–15

Plant	Pre–sowing	Temp °F / °C	Cover	Germination
Echinops		64–75 / 18–24	No	14–21
Echium		64–72 / 18–22	Lightly	7–21
Elaeagnus	Stratify 12 weeks	68–77/ 20–25	Yes	28–42
Elaeocarpus		59–72 / 15–22	Lightly	21–42
Entelea		64–77 / 18–25	Lightly	10–20
Epacris		59–72 / 15–22	Lightly	21–35
Ephedra	Perhaps stratify	64–72 / 18–22	Lightly	14_35
Eranthis	Stratify 6 weeks	59–64 / 15–18	Lightly	14–35
Eremurus		55–72 / 13–22	Yes	14–35
Erica		59–72 / 15–22	No	15–35
Erigeron		55–64 / 13–18	Lightly	14–21
Erinus		64–77 / 18–25	Lightly	14–28
Eriobotrya	Soak 24 hours	64–75 / 18–24	Yes	10–28
Eriogonum		64–72 / 18–22	Lightly	14–28
Erodium		59–75 / 15–24	Lightly	12–28
Eruca		59–70 / 15–21	Lightly	5–10
Eryngium		64–77 / 18–25	Lightly	5–10
Erysimum		55 64 / 13–18	Lightly	5–15
Erythrina	Scarify & soak 24 hours	68 77/ 20–25	Yes	7–21
Erythronium	Better stratified 8 weeks	55–68 / 13–20	Lightly	14–35
Eschscholzia		64–72 / 18–22	Lightly	5–10
Eucalyptus		64–77 / 18–25	Lightly	14–28
Eucomis		68–77/ 20–25	Yes	14–28
Eugenia		64–75 / 18–24	Lightly	14–28
Eupatorium	Perhaps stratify 4 weeks	59–72 / 15–22	Lightly	10–28
Euphorbia – garden forms		64 72 / 18–22	No	5–15
Eustoma		59–72 / 15–22	No	10–35
Exacum		68–77/ 20–25	Yes	14–21
Exochorda	Stratify 6 weeks	64–72 / 18–22	Yes	14–28
Fagopyrum		59–77 / 15–25	Lightly	14–21
Fagus – grafting stock	Stratify 12 weeks	59–72 / 15–22	Lightly	14–35
Fallopia		59–68 / 15–20	Yes	14–21
Fatsia		64–77 / 18–25	No	28–40
Felicia		57–70 / 14–21	Lightly	21–35
Ficus		68–82 / 20–28	No	14–21
Filipendula		59–72 / 15–22	Lightly	14–35
Foeniculum		64–72 / 18–22	Yes	7–14
Fragaria		61–68 / 16–20	Lightly	21–28
Francoa		59–68 / 15–20	Lightly	7–21
Franklinia	Stratify 4–6 weeks	68–77/ 20–25	Yes	21–35
Fraxinus	Stratify 12 weeks	59–72 / 15–22	Lightly	14–35
Freesia	Soak 12 hours	64–72 / 18–22	Yes	21–28
Fremontodendron	Scarify or soak 24 hours	64–75 / 18–24	Yes	21–42+
Fritillaria	Stratify 8 weeks	59–70 / 15–21	Lightly	14–35
Fuchsia		68–77/ 20–25	No	21–35
Gaillardia		68–75 / 20–24	No	5–15
Galanthus	Better stratified 6 weeks	55–68 / 13–20	Yes	14–42
Galega		59–72 / 15–22	Lightly	7–21

Plant	Pre–sowing	Temp °F / °C	Cover	Germination
Galtonia		64–72 / 18–22	Yes	14–21
Gardenia		68–77/ 20–25	Yes	21–35
Gaultheria		55–72 / 13–22	No	21–56+
Gazania		64–75 / 18–24	Yes	7–14
Geissorhiza		59–75 / 15–24	Lightly	14–28
Geniostoma		59–75 / 15–24	Lightly	14–35
Gentiana	Stratify 4–6 weeks	64–77 / 18–25	Lightly	14–28
Geranium		59–72 / 15–22	Lightly	14–42
Gerbera		68–77/ 20–25	Yes	10–15
Geum	Perhaps stratify 4 weeks	64–72 / 18–22	No	8–20
Ginkgo	Stratify 8–12 weeks	68–77/ 20–25	Yes	28–35
Gladiolus		64–77 / 18–25	Yes	21–42
Glaucium		59–75 / 15–24	Lightly	14–28
Gleditsia	Scarify & soak 24 hours	64–75 / 18–24	Yes	7–21
Globularia	Perhaps stratify 4 weeks	55–64 / 13–18	Lightly	7–14
Gloriosa		68–77/ 20–25	Yes	28–35
Gomphrena		68–77/ 20–25	Lightly	10–15
Goodia	Scarify & soak 24 hours	64–75 / 18–24	Yes	7–21
Gordonia		59–72 / 15–22	Lightly	14–35
Grevillea	Soak 12 hours	68–82 / 20–28	Yes	14–35
Gunnera		59–68 / 15–20	Yes	14–35
Gymnocladus	Stratify 8–10 weeks	59–72 / 15–22	Yes	14–42
Gypsophila		68–77/ 20–25	No	5–15
Haemanthus		64–77 / 18–25	Lightly	20–80
Hakea	Scarify & soak 12 hours	64–77 / 18–25	Yes	14–28
Halesia	Better stratified 8 weeks	59–72 / 15–22	Lightly	10–28
Hamamelis	Stratify 12 weeks	59–72 / 15–22	Lightly	21–42+
Hardenbergia	Soak 8 hours	64–75 / 18–24	Yes	5–21
Harpephyllum	Remove all fruit pulp	64–75 / 18–24	Lightly	10–28
Hebe		55–75 / 13–24	Lightly	14–28
Hedycarya		59–72 / 15–22	Lightly	14–35
Hedychium		68–77/ 20–25	Yes	14–28
Helenium		68–72 / 20–22	No	7–14
Helianthemum		68–77/ 20–25	Lightly	14–21
Helianthus		64–75 / 18–24	Yes	5–10
Helichrysum		68–75 / 20–24	No	7–14
Heliophila		59–72 / 15–22	Lightly	10–21
Heliopsis		64–72 / 18–22	Lightly	5–15
Heliotropium		64–75 / 18–24	Lightly	10–21
Helipterum		68–75 / 20–24	Lightly	5–15
Helleborus	Stratify 12 weeks or sow outdoors	64–77 / 18–25	Yes	30–300+
Hemerocallis	Stratify 6 weeks	59–72 / 15–22	Yes	21–56
Hermodactylus	Better stratified 4 weeks	59–72 / 15–22	Lightly	10–28
Herpolirion		57–68 / 14–20	Lightly	10–28
Hesperis		64–77 / 18–25	No	5–10
Heuchera		59–70 / 15–21	No	20–30
Hibiscus	Soak 12 hours	68–77/ 20–25	Yes	10–15
Hippeastrum		68–77/ 20–25	Lightly	28–42

Plant	Pre–sowing	Temp °F / °C	Cover	Germination
Hippophae	Stratify 6–8 weeks	59–68 / 15–20	Yes	21–42
Hoheria		59–72 / 15–22	Lightly	14–35
Hosta	Perhaps stratify 6 weeks	64–72 / 18–22	Lightly	14–21
Hoya		68–77/ 20–25	Lightly	8–21
Humulus		68–77/ 20–25	Lightly	21–35
Hyacinthoides		59–68 / 15–20	Lightly	14–35
Hyacinthus		64–72 / 18–22	Lightly	21–35
Hymenanthera		59–72 / 15–22	Lightly	14–35
Hymenosporum		64–77 / 18–25	Yes	14–35
Hypericum		59–70 / 15–21	Yes	10–21
Hypoestes		68–77/ 20–25	Lightly	7–14
Hyssopus		59–70 / 15–21	Lightly	7–14
Iberis – annual		64–75 / 18–24	Lightly	7–15
Iberis – perennial		59–64 / 15–18	No	14–21
Idesia	Stratify 6 weeks	59–72 / 15–22	Lightly	10–28
Ilex	Stratify 12 weeks	64–72 / 18–22	Yes	30–180+
Impatiens balsamina		64–72 / 18–22	Lightly	7–11
Impatiens wallerana		72–77 / 22–25	Lightly	10–18
Incarvillea		59–64 / 15–18	Lightly	21–35
Indigofera	Scarify & soak 24 hours	64–75 / 18–24	Yes	10–28
Inula		64–72 / 18–22	Lightly	7–21
Ipomoea	Scarify or soak 12 hours	64–75 / 18–24	Yes	5–10
Iris	Stratify 6–8 weeks	59–75 / 15–24	Lightly	21–50
Isopogon	Gently scarify	64–75 / 18–24	Lightly	21–42+
Ixia		64–75 / 18–24	Lightly	14–35
Ixiolirion		64–75 / 18–24	Lightly	14–35
Jacaranda	Better soaked 8 hours	64–82 / 18–28	Yes	7–15
Jasione		64–72 / 18–22	Lightly	10–15
Jasminum		68–77/ 20–25	Lightly	14–35
Jeffersonia	Stratify 6 weeks	59–68 / 15–20	Yes	21–42
Jovellana		55–72 / 13–22	Lightly	7–21
Juglans	Stratify 12 weeks	59–72 / 15–22	Yes	14–35+
Justicia		64–72 / 18–22	No	14–35
Kalanchoe		64–75 / 18–24	Lightly	10–15
Kalmia	Stratify 12 weeks	64–72 / 18–22	Lightly	21–42
Kalmiopsis		57–68 / 14–20	No	21–35
Knautia	Perhaps stratify	64–72 / 18–22	Lightly	14–28
Knightia	Gently scarify	59–75 / 15–24	Lightly	21–42+
Kniphofia		64–75 / 18–24	No	21–28
Koelreuteria	Soak for 8 hours	64–72 / 18–22	Yes	21–35
Kunzea		59–75 / 15–24	Lightly	10–35
Laburnum	Scarify or soak 24 hours	59–70 / 15–21	Yes	28–70
Lachenalia		64–75 / 18–24	Lightly	14–28
Lactuca		64–72 / 18–22	Lightly	5–10
Lagerstroemia		68–77/ 20–25	Yes	14–18
Lantana		68–77/ 20–25	Yes	35–60
Lapageria		64–75 / 18–24	Lightly	21–42
Lapeirousia		59–75 / 15–24	Lightly	10–28

Plant	Pre–sowing	Temp °F / °C	Cover	Germination
Larix	Better stratified 8 weeks	59–72 / 15–22	Lightly	14–35
Lathyrus – perennial	Scarify or soak 24 hours	55–64 / 13–18	Yes	10–21
Lathyrus	Scarify or soak 24 hours	59–68 / 15–20	Yes	7–15
Laurelia		59–72 / 15–22	Lightly	21–42
Lavandula	Stratify 4 weeks	64–75 / 18–24	No	14–21
Lavatera		64–72 / 18–22	Yes	5–10
Leontopodium	Perhaps stratify 4 weeks	64–72 / 18–22	No	15–21
Leonurus		64–75 / 18–24	Lightly	10–28
Leptospermum		59–75 / 15–24	Lightly	10–35
Leschenaultia		64–75 / 18–24	Lightly	20–50+
Leucadendron	Gently scarify or soak	64–75 / 18–24	Yes	21–42+
Leucojum		59–68 / 15–20	Lightly	21–42
Leucopogon		55–68 / 13–20	Lightly	21–60+
Leucospermum	Gently scarify or soak	64–75 / 18–24	Yes	21–42+
Leucothoe		59–72 / 15–22	No	14–35+
Levisticum		59–70 / 15–21	Lightly	7–14
Lewisia	Stratify 4 weeks	64–72 / 18–22	Lightly	28–50+
Liatris		64–70 / 18–21	No	21–28
Libertia		59–68 / 15–20	Yes	14–28
Libocedrus	Perhaps stratify 4 weeks	55–68 / 13–20	Yes	21–42+
Lilium	Some need stratifying	64–72 / 18–22	Yes	21–42
Limnanthes		59–72 / 15–22	Lightly	7–21
Limonium		64–72 / 18–22	Yes	5–14
Linaria		55–59 / 13–15	Lightly	10–15
Lindera	Stratify 6 weeks	59–77 / 15–25	Lightly	30–60
Linum		64–72 / 18–22	Lightly	14–28
Liquidambar	Stratify 8–12 weeks	59–72 / 15–22	Yes	14–35
Liriodendron	Stratify 8–12 weeks	59–72 / 15–22	Lightly	21–42+
Liriope	Soak 24 hours	61–72 / 16–22	Yes	28–35
Lithops		68–77/ 20–25	Lightly	10–15
Littonia		64–75 / 18–24	Yes	14–35
Lobelia – annual		68–77/ 20–25	Lightly	14–21
Lobelia – perennial	Stratify 12 weeks	61–68 / 16–20	Lightly	14–21
Lobularia		64–77 / 18–25	No	7–10
Lomatia		59–72 / 15–22	Yes	20–50+
Luculia		64–72 / 18–22	Lightly	14–21
Lunaria		64–75 / 18–24	No	10–15
Lupinus	Scarify or soak 24 hours	64–75 / 18–24	Yes	5–15
Lychnis		64–72 / 18–22	No	21–28
Lycopersicon		70–77 / 21–25	Yes	7–14
Lycoris	Better stratified 6 weeks	59–72 / 15–22	Lightly	10–28
Lysichiton		55–68 / 13–20	Lightly	14–35
Lythrum		64–72 / 18–22	Lightly	14–21
Macadamia	Scarify & soak 48 hours	64–77 / 18–25	Yes	14–35+
Magnolia	Stratify 12–16 weeks	64–72 / 18–22	Yes	30–90
Mahonia	Better stratified 8 weeks	59–72 / 15–22	Yes	21–42
Malcolmia		57–68 / 14–20	Lightly	7–21
Malus	Stratify 12 weeks	59–72 / 15–22	Yes	14–35

Plant	Pre–sowing	Temp °F / °C	Cover	Germination
Malva		64–72 / 18–22	Yes	5–15
Mammillaria		64–72 / 18–22	Lightly	14–35
Mandevilla	Soak 8 hours	68–77/ 20–25	Yes	14–35
Marrubium		64–72 / 18–22	Yes	10–15
Matricaria		55–64 / 13–18	Lightly	7–14
Matthiola		64–75 / 18–24	No	7–14
Meconopsis		59–68 / 15–20	Lightly	14–28
Melaleuca		64–75 / 18–24	Lightly	14–35
Melicytus		59–72 / 15–22	Lightly	14–28
Melissa		64–72 / 18–22	No	10–15
Mentha		64–72 / 18–22	Lightly	10–15
Menziesia	Stratify 6–8 weeks	59–68 / 15–20	No	21–42
Meryta		64–77 / 18–25	Yes	21–42
Mesembryanthemum		64–72 / 18–22	No	7–15
Mespilus	Stratify 6–8 weeks	59–68 / 15–20	Yes	21–42
Metasequoia	Stratify 8–12 weeks	59–72 / 15–22	Yes	21–42+
Metrosideros		59–77 / 15–25	Lightly	14–35
Mimulus		59–70 / 15–21	No	5–10
Mirabilis		64–72 / 18–22	Yes	5–10
Moluccella		59–75 / 15–24	No	12–21
Monarda		59–72 / 15–22	Lightly	14–21
Monstera		75–82 / 24–28	Lightly	14–21
Morus		64–75 / 18–24	Yes	21–35+
Muehlenbeckia		55–68 / 13–20	Lightly	10–28
Musa	Remove all fruit pulp	68–81 / 20–27	Lightly	10–28
Muscari		59–68 / 15–20	Lightly	14–28
Myoporum		55–72 / 13–22	Lightly	10–35
Myosotidium		55–64 / 13–18	Yes	14–35+
Myosotis – annual		55–68 / 13–20	Yes	14–28
Myosotis – perennial		64–72 / 18–22	No	7–14
Myrtus		59–77 / 15–25	Lightly	14–35
Nelumbo	Scarify then sprout in water	64–75 / 18–24	No	10–30+
Nemesia		61–72 / 16–22	Yes	8–14
Nemophila		61–72 / 16–22	Yes	10–15
Nepeta		64–72 / 18–22	Yes	5–15
Nerine		59–72 / 15–22	Lightly	14–42
Nertera		59–70 / 15–21	Yes	14–28
Nestegis		59–77 / 15–25	Yes	21–42
Nicotiana		70–75 / 21–24	No	10–18
Nierembergia		68–77 / 20–25	Lightly	5–15
Nigella		64–72 / 18–22	Lightly	7–15
Nomocharis		59–72 / 15–22	Lightly	14–35
Nothofagus	Better stratified 8 weeks	55–68 / 13–20	Lightly	14–42
Notospartium	Scarify or soak 24 hours	59–72 / 15–22	Yes	10–28
Nyssa	Stratify 8–12 weeks	59–72 / 15–22	Yes	21–42+
Ochna	Better soaked 24 hours	64–77 / 18–25	Yes	7–21
Ocimum		64–75 / 18–24	Lightly	5–10
Oenothera		68–77/ 20–25	No	7–18

Plant	Pre–sowing	Temp °F / °C	Cover	Germination
Oldenburgia		64–75 / 18–24	Lightly	10–28
Olearia	Seed must be fresh	55–72 / 13–22	Lightly	14–42
Omphalodes		59–72 / 15–22	Lightly	10–20
Ophiopogon	Soak & stratify 4 weeks	59–72 / 15–22	Yes	10–28
Opuntia	Soak 8 hours	64–77 / 18–25	Lightly	14–35
Origanum		64–72 / 18–22	Lightly	5–10
Ornithogalum		59–75 / 15–24	Lightly	14–35
Orphium		59–75 / 15–24	Lightly	10–21
Orthrosanthus		64–75 / 18–24	Lightly	14–28
Ostrya	Stratify at least 6 weeks	64–75 / 18–24	Lightly	21–42
Ourisia		55–72 / 13–22	Lightly	10–21
Oxydendrum		59–75 / 15–24	Lightly	14–35+
Oxylobium	Scarify & soak 24 hours	64–75 / 18–24	Yes	7–21
Pachystegia		55–68 / 13–20	Lightly	14–28
Paeonia	Sprout then stratify 8 weeks	64–72 / 18–22	Yes	30–60
Panax		64–75 / 18–24	Lightly	10–28
Papaver		64–75 / 18–24	No	5–15
Papaver orientale		61–72 / 16–22	No	10–15
Parahebe		55–72 / 13–22	No	7–28
Paraserianthes	Soak 24 hours	64–75 / 18–24	Yes	7–28
Paratrophis		64–77 / 18–25	Lightly	14–35
Parrotia	Stratify 8 weeks	59–72 / 15–22	Lightly	21–60+
Parsonsia		59–77 / 15–25	Yes	14–42
Passiflora		68–77/ 20–25	Lightly	28–45
Pastinaca	Better soaked 24 hours	59–72 / 15–22	Lightly	14–35
Paulownia	Sow fresh or stratify if stored	64–75 / 18–24	No	10–35+
Pelargonium	Scarify	64–77 / 18–25	Yes	10–28
Penstemon		55–68 / 13–20	Yes	14–35
Pentas		68–75 / 20–24	No	5–15
Pericallis		70–75 / 21–24	No	10–15
Perovskia		64–72 / 18–22	Lightly	14–35
Persea		68–81 / 20–27	Partly	21–50+
Persoonia	Gently scarify	64–75 / 18–24	Lightly	14–35+
Petroselinum	Better soaked 12 hours	64–72 / 18–22	Yes	14–21
Petunia		72–81 / 22–27	No	7–14
Phacelia		55–72 / 13–22	Lightly	7–21
Phaseolus		64–72 / 18–22	Yes	7–14
Phebalium		59–72 / 15–22	Lightly	14–35
Phellodendron	Stratify 6–8 weeks	59–68 / 15–20	Yes	21–42
Philadelphus	Stratify 4–6 weeks	59–64 / 15–18	Yes	21–42
Phlomis		55–72 / 13–22	Yes	14–21
Phlox – perennial	Stratify 4 weeks	64–72 / 18–22	Yes	21–35
Phlox drummondii		59–64 / 15–18	Yes	10–15
Phoenix	Scarify & soak 8 hours	64–77 / 18–25	Yes	10–35
Phormium		55–77 / 13–25	Yes	10–35
Phygelius		68–77/ 20–25	Lightly	10–15
Phylica	Remove all chaff	61–75 / 16–24	Lightly	10–28
Phyllocladus		55–68 / 13–20	Yes	21–60+

Plant	Pre–sowing	Temp °F/°C	Cover	Germination
Phyllodoce		59–68 / 15–20	No	14–35
Phyllostachys		64–75 / 18–24	No	14–35
Physalis		59–70 / 15–21	No	7–14
Physostegia		64–70 / 18–21	Lightly	7–14
Phyteuma		59–72 / 15–22	Lightly	14–28
Phytolacca	Perhaps stratify	59–72 / 15–22	Yes	14–35
Picea	Stratify 12 weeks	55–68 / 13–20	Yes	21–42
Pieris		59–72 / 15–22	No	15–35
Pimelia		59–72 / 15–22	Lightly	14–28
Pimpinella		64–72 / 18–22	Lightly	10–15
Pinus	Some need stratifying	59–72 / 15–22	Yes	14 42
Pisonia		59–77 / 15–25	Yes	14–35
Pisum		59–72 / 15–22	Yes	7–14
Pittosporum		59–72 / 15–22	Yes	14–42+
Plagianthus		55–72 / 13–22	Lightly	14–42
Platycodon		59 70 / 15–21	No	7–14
Plumbago		68–77/ 20–25	Lightly	21–35
Podalyria	Soak 24 hours	64–75 / 18–24	Yes	14–35
Podocarpus	Better stratified 8 weeks	59–72 / 15–22	Yes	21–60+
Podolepis		61–72 / 16–22	Lightly	10–28
Podophyllum	Stratify 6 weeks	59–68 / 15–20	Yes	21–42
Polemonium		68–77/ 20–25	Lightly	18–28
Polygonum		68–77/ 20–25	Lightly	18–28
Pomaderris		59–77 / 15–25	Lightly	14–28
Portulaca		72–82 / 22–28	No	7–14
Potentilla	Some need stratifying	59 72 / 15–22	Lightly	14 21
Primula	Stratify alpines	59–72 / 15–22	No	14–28
Protea	Gently scarify or soak	64–75 / 18–24	Yes	15–42
Prunus	Stratify 12 weeks	59–75 / 15–24	Yes	21–42+
Pseudopanax	Seed must be fresh	59–72 / 15–22	Yes	14–35
Pseudowintera	Better soaked 8 hours	55–72 / 13–22	Yes	14–42
Psidium	Remove all fruit pulp	68–77/ 20–25	Lightly	10–28
Ptelea	Stratify 8–12 weeks	64–72 / 18–22	Yes	20–50
Pterocarya		64–72 / 18–22	Yes	21–42
Pterostyrax	Better stratified 8 weeks	64–75 / 18–24	Yes	14–35
Pulmonaria		55–72 / 13–22	Lightly	10–28
Pulsatilla		55–68 / 13–20	Lightly	14–21
Punica		68–77/ 20–25	No	/ 20–28
Puya	Better stratified 4 weeks	55–72 / 13–22	Lightly	14–35
Pyrus	Stratify 12 weeks	59–72 / 15–22	Yes	14–35
Quercus	Stratify 12 weeks	59–72 / 15–22	Yes	15–42
Quintinia		59–72 / 15–22	Lightly	14–35
Quisqualis		68–77/ 20–25	Lightly	20–50
Ranunculus		59–64 / 15–18	Lightly	14–21
Raphanus		55–70 / 13–21	Lightly	3–7
Ratibida		64–72 / 18–22	Lightly	14–28
Rehmannia		68–77/ 20–25	Lightly	14–21
Reseda		68–77/ 20–25	No	5–10

Plant	Pre–sowing	Temp °F / °C	Cover	Germination
Rhabdothamnus		64–77 / 18–25	No	14–35
Rhododendron		59–72 / 15–22	Lightly	21–60
Rhodotypos	Stratify 8–12 weeks	59–68 / 15–20	Yes	20–50
Rhopalostylis	Soak 24 hours	64–77 / 18–25	Yes	21–60+
Rhus	Stratify 8 weeks	59–72 / 15–22	Lightly	21–42+
Ricinus	Better if soaked 8 hours	68–77 / 20–25	Yes	14–21
Robinia	Soak then stratified 8 weeks	59–72 / 15–22	Yes	14–35
Rodgersia	Better stratified 6 weeks	55–68 / 13–20	Lightly	10–28
Romneya		64–72 / 18–22	Lightly	14–35
Romulea		64–75 / 18–24	Yes	21–42+
Rosa	Stratify 8–12 weeks	55–72 / 13–22	Yes	21–28
Rosmarinus		64–72 / 18–22	No	10–21
Rubia		57–68 / 14–20	Yes	14–28
Rudbeckia		68–75 / 20–24	Lightly	5–10
Rumex		55–68 / 13–20	Lightly	7–14
Ruscus	Stratify 12 weeks	59–77 / 15–25	Yes	21–42+
Ruta		59–72 / 15–22	Lightly	7–14
Sagina		55–64 / 13–18	Lightly	10–15
Salpiglossis		68–75 / 20–24	Lightly	14–21
Salvia		64–75 / 18–24	Lightly	5–15
Salvia officinalis – culinary		64–72 / 18–22	Yes	5–15
Sambucus		64–82 / 18–28	No	5–15
Sandersonia	Keep seed mix moist	64–75 / 18–24	Lightly	21–60+
Santolina		61–72 / 16–22	Lightly	14–21
Saponaria		64–72 / 18–22	Yes	5–10
Sarcococca	Better stratified 8 weeks	59–72 / 15–22	Yes	21–42+
Satureja		64–72 / 18–22	No	10–20
Saxifraga		61–77 / 16–25	Lightly	14–21
Scabiosa		64–72 / 18–22	No	10–20
Schefflera		68–77 / 20–25	Yes	14–28
Schinus		64–77 / 18–25	Lightly	14–35
Schizanthus		59–72 / 15–22	No	7–14
Schotia	Soak 24 hours	64–75 / 18–24	Yes	10–28
Sciadopitys	Stratify 8 weeks	59–75 / 15–24	Yes	14–42
Scutellaria		59–75 / 15–24	Lightly	14–35
Sedum		68–82 / 20–28	No	5–15
Sempervivum		73–82 / 23–28	Lightly	14–35
Senecio		59–77 / 15–25	Lightly	10–15
Senna	Scarify	68–82 / 20–28	Yes	5–14
Sequoiadendron	Better stratified 8 weeks	59–72 / 15–22	Yes	21–42+
Serissa		64–77 / 18–25	Lightly	10–21
Serruria		64–75 / 18–24	Yes	14–42
Sesamum		64–75 / 18–24	Lightly	4–10
Silene		64–72 / 18–22	Lightly	14–21
Sinningia		64–73 / 18–23	No	14–21
Sisyrinchium		61–75 / 16–24	Yes	10–28
Solanum		64–72 / 18–22	No	7–14
Solanum – Eggplant		68–75 / 20–24	Lightly	7–14

Plant	Pre–sowing	Temp °F / °C	Cover	Germination
Solenostemon (Coleus)		68–77/ 20–25	No	10–15
Solidago	Better stratified 8 weeks	55–68 / 13–20	Lightly	14–28
Sophora	Scarify or soak 48 hours	59–77 / 15–25	Yes	14–42
Sorbus	Stratify 14 weeks	55–72 / 13–22	Lightly	18–35
Sparaxis	Seed must be fresh	59–75 / 15–24	Lightly	14–35
Spartium	Scarify or soak 24 hours	59–72 / 15–22	Yes	7–21
Sphaeralcea		64–75 / 18–24	Yes	14–35
Spinacia	Perhaps stratify 2 weeks	50–59 / 10–15	Lightly	7–14
Stachys		64–72 / 18–22	No	5–15
Stauntonia	Remove all fruit pulp	59–72 / 15–22	Yes	15–42+
Stenocarpus	Gently scarify or soak	64–79 / 18–26	Yes	14–35+
Stephanotis		68–77/ 20–25	Lightly	28–42
Stokesia		64–72 / 18–22	Lightly	21–35
Strelitzia	Soak up to 72 hours	79–90 / 26–32	Lightly	30–180+
Streptocarpus		72–79 / 22–26	No	14–21
Styrax	Stratify 6 weeks	64–77 / 18 25	Yes	14–35
Tagetes		68–77/ 20–25	Lightly	5–10
Tanacetum		59–72 / 15 22	Lightly	7–15
Taxodium	Stratify 12 weeks	59–77 / 15–25	Yes	21–42
Taxus	Stratify 12 weeks	64–72 / 18–22	Yes	30–250+
Tecomanthe		64–77 / 18–25	Yes	14–35
Tecophilea		64–77 / 18–25	Yes	14–35
Telopea	Gently scarify or soak	64–75 / 18–24	Yes	21–35
Tetrapanax		59–72 / 15–22	Yes	14–35+
Tetrapathea	Remove all fruit pulp	64–75 / 18–24	Lightly	10–28
Teucrium		64–72 / 18 22	Lightly	21–35
Thalictrum		64–72 / 18–22	Lightly	14–35
Thryptomene		64–75 / 18–24	Lightly	14–35
Thymus		64–72 / 18–22	Yes	5–15
Tiarella		59–72 / 15–22	Lightly	14–28
Tigridia		64–77 / 18–25	Yes	18–28
Tilia	Stratify 12 weeks	57–72 / 14–22	Yes	14–35
Torenia		68–75 / 20–24	No	7–15
Toronia		64–77 / 18–25	Lightly	14–35
Townsendia		59–72 / 15–22	Lightly	14–28
Trachymene		64–72 / 18–22	Yes	14–21
Tradescantia		64–72 / 18–22	Lightly	21–35
Trigonella		57–72 / 14–22	Lightly	4–10
Trillium	Stratify 10 weeks (twice)	59–70 / 15–21	Yes	30–150+
Tropaeolum		64–72 / 18–22	Yes	10–15
Tsuga	Stratify 8 weeks	59–72 / 15–22	Yes	20–50
Tulipa	Stratify 8 weeks	59–64 / 15–18	Yes	14–21
Tweedia		64–72 / 18–22	Yes	10–15
Ulmus	Stratify 12 weeks	59–72 / 15–22	Yes	18–42
Umbellularia		64–75 / 18–24	Lightly	20–40
Vaccinium		59–68 / 15–20	No	20–50
Valeriana		64–72 / 18–22	Lightly	21–28
Veltheimia	Seed must be fresh	64–75 / 18–24	Lightly	10–28

73

Plant	Pre–sowing	Temp °F / °C	Cover	Germination
Verbascum		68–75 / 20–24	No	5–15
Verbena		70–77 / 21–25	Lightly	5–15
Veronica		64–75 / 18–24	No	7–14
Viburnum	Stratify 10 weeks (twice)	64–72 / 18–22	Yes	42–90+
Viguiera		59–75 / 15–24	Lightly	10–28
Vinca		68–77 / 20–25	Yes	14–21
Viola		64–75 / 18–24	Lightly	7–15
Virgilia	Soak 24 hours	59–77 / 15–25	Yes	10–28
Vitex		64–77 / 18–25	Yes	14–42
Wachendorfia		64–77 / 18–25	Yes	14–42
Weinmannia		59–72 / 15–22	Lightly	21–42
Wisteria	Scarify & soak 24 hours	55–64 / 13–18	Yes	28–42
Xeranthemum		64–72 / 18–22	Lightly	10–15
Xeronema		64–75 / 18–24	Lightly	14–35
Yucca		61–77 / 16–25	Yes	18–28
Zantedeschia		68–77/ 20–25	No	28–42
Zea		68–75 / 20–24	Yes	5–10
Zelkova	Stratify 12 weeks	57–72 / 14–22	Yes	21–42+
Zephyranthes		59–75 / 15–24	Lightly	14–42
Zinnia		68–75 / 20–24	Lightly	5–10

5
Propagation by division

Plants that form large clumps with multiple growth points, such as many of the herbaceous perennials, can be propagated by division. The method involves nothing more than breaking up established clumps into smaller pieces. Provided each division has a growth point and a few roots, it should be capable of surviving as an independent plant.

Division is a simple method well suited to the open garden. It requires no specialized equipment and produces large, nearly mature plants. However, because it involves significant disturbance of established plants, and as many divisible perennials will also grow from cuttings, you may prefer to try that method when you want something less disruptive.

Rosettes, runners and offsets

Rosettes

The plants that divide most easily are those that form foliage rosettes, such as *Ajuga* and the saxifrages. They break up neatly into new plants, each with a rosette of foliage and some roots. This simple separation can, in most cases, be done at any time of year.

Runners

Runners are side-shoots that spread along the soil surface. In many cases, small plantlets will grow from these runners. The plantlets will tend to strike roots wherever they come in contact with the soil and they can be removed from the parent plant and treated as individuals.

Offsets

Many cacti and succulents produce side-shoots that separate from the parent plant by themselves. These are known as offsets and all you need to do to produce a new plant is to pot them. Alternatively, you can remove the offsets before they separate. They will have no roots, but pot them and the roots will soon develop. It is often a good idea to leave potting very succulent offsets for a few hours after removing them from the parent plant so that the damaged tissue has a chance to dry and form a callus.

Early spring growth of *Monarda* 'Prairie Night.' The perennial bergamots form large clumps that divide readily.

Mammillaria decipiens, a clustering cactus. The offsets can be potted and grown on.

Herbaceous perennials

Deciduous herbaceous perennials, such as *Hosta*, *Lythrum* and *Phlox*, are best divided near the end of their dormant period or when just starting into growth. The growth points are easier to identify then and the plants can begin to grow as soon as they have been divided. If you divide in autumn, when the plants first become dormant, they have to endure the entire winter before starting to grow. Getting the divisions growing and the wounds healed as quickly as possible is the best way to avoid the soft rots that occur on cut surfaces. Dusting the wounds with a fungicide before replanting also helps.

Evergreen and semi-deciduous perennials often have distinct foliage clusters and heavy fibrous-rooted crowns. Many lily family plants, such as New Zealand flax (*Phormium*) and *Libertia*, fall into this category. They, too, are best divided in late winter or early spring, but they may need to be planted in nursery beds until established because they often have very few roots per division.

Natural layering and aerial roots

Natural layering

Natural layers form when a stem is kept in contact with the soil for a prolonged period. Many perennials and shrubs, particularly ground covers such as thyme (*Thymus*), routinely form natural layers that can be removed and grown on.

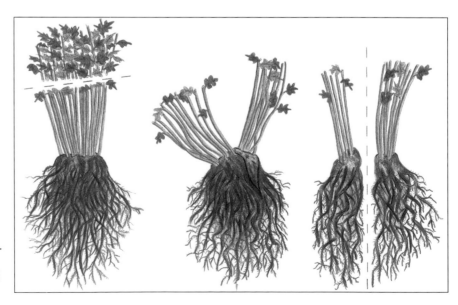

Dividing perennials: Clump-forming fibrous-rooted perennials can be divided by simply breaking the plant into rooted divisions and then replanting.

Aerial roots

Some plants, usually tropical in origin or native to very humid areas, produce aerial roots that grow directly from their stems, often some distance above ground. A few climbers, such as ivy (*Hedera*) and *Campsis*, have aerial roots that are modified into grasping tools that support the plants as they climb. If a stem with aerial roots is cut off and planted, the roots often develop as normal subterranean roots. However, this is not always successful and is best done under greenhouse conditions.

A self-layered piece of prostrate rosemary (*Rosamarinus officinalis* 'Lavandulaceus'). This well-rooted piece can be potted and grown on.

Suckers

Suckers are strong growing shoots that emerge from the base of some trees and shrubs. Plants that produce suckers include lilacs (*Syringa*), elms (*Ulmus*), toon (*Toona*) and flowering quince (*Chaenomeles*).

Often, the suckers can be removed with roots attached and grown on. Suckers of deciduous plants should be removed as early as possible in the growing season to ensure that they have adequate time to become established before dropping their foliage in autumn.

Rhizomes, tubers, corms and bulbs

Rhizomes, tubers, corms and bulbs are stems modified into food storage organs. Plants that produce them can usually withstand an extended period of dormancy, during which they survive by using their stored reserves. Rhizomes, tubers and corms are capable of being divided and some bulbs can be separated into scales that can be grown on.

Algerian ivy (*Hedera canariensis*) with developing aerial roots. When used as cuttings, those on the softer wood (top) establish better than those on woody stems (lower).

Small suckers surround the base of a Cappadocium or Caucasian maple (*Acer cappadocian*).

Rhizomes

Rhizomes grow on or just below the soil surface. They are segmented stems with buds at the nodes, just like an above-ground stem. The difference is that, in addition to foliage, there are roots at each of the nodes. As each node has both a leaf bud and a root bud, propagation is simply a matter of breaking up the rhizome at the nodes. Mints and the creeping grasses, such as kikuyu and couch, are common rhizomatous plants.

Breaking up rhizomes is a simple matter, but the divisions are often lacking in true roots and foliage, which makes them less able to support themselves than, say, a strong-growing herbaceous perennial division. For this reason it is better to err on the generous side with your divisions. Dust any cut surface with a fungicide—powdered sulfur is suitable. Water and feed the fresh plants until they are growing well. Like succulents, very fleshy rhizomes are better left to dry for a while before planting, to lessen the risk of rotting.

The different structures of food-bearing organs.

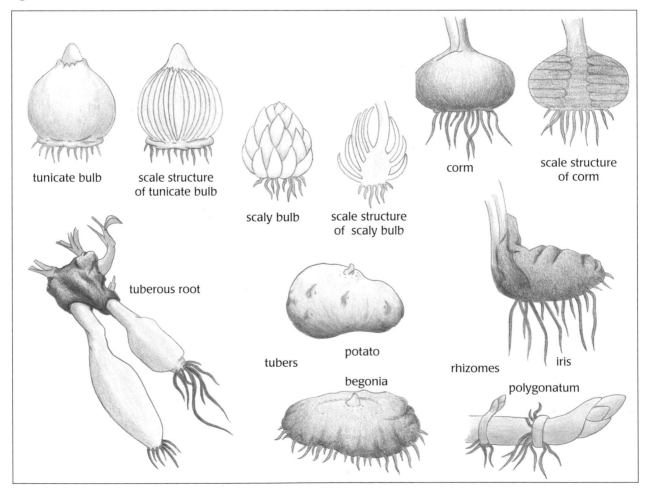

tunicate bulb

scale structure
of tunicate bulb

scaly bulb

scale structure
of scaly bulb

corm

scale structure
of corm

tuberous root

tubers

potato

begonia

rhizomes

iris

polygonatum

Stolons are similar to rhizomes except that they tend to produce new plantlets only at their tips rather than also at the nodes. The tip plantlets can be removed and grown on, which often causes the stolons to form more plantlets along their length.

Tubers

Tubers are always found underground at the base of the main stems of the plant. They are usually swollen and are easily removed from the fibrous root system. The tubers form during the growing season, at the end of which the top growth dies away, leaving the tubers to sprout, grow and repeat the cycle the following season. The most common tuber is undoubtedly the potato, and it is typical of the general style of growth.

Propagating tuberous plants is just a matter of digging up the tubers at the end of the growing season and separating them from the crown. They usually break away without any difficulty, but if they are stubborn let them dry a little, then carefully prize them apart. If you have very well-drained soil the tubers may be planted out right away, otherwise store them in barely moist sawdust in a frost-free place and plant them out in spring.

Most tubers have several growth points or eyes. It is possible to cut a tuber into smaller pieces, each with an eye. The best time for this is in early spring, just as the eyes are starting to swell. Because there is always the possibility that an eye may be damaged or may fail to develop, it is best to keep to at least two eyes per division.

Each cut piece will have a large area of exposed flesh, making it very prone to fungal diseases and rotting. Dip the cut sections of tuber in a fungicide for protection—powdered sulfur is safe and easy to use. Unless the tuber is quite shriveled, do not plant the pieces right away; instead leave them in a warm, dry place out of direct sun for a day or so while the cut surfaces dry and calluses form.

Corms

Corms and bulbs look similar to one another but there are significant differences between them. While these differences are not very important in terms of cultivation there are special methods of propagation for each type.

Whereas bulbs are composed of layers of tissue called scales and increase in girth each year by adding more layers of scales, most corms produce a completely new structure each year. The old corm withers away while a new and larger corm forms on top. Often small corms, or cormels, form between the old and new corms. These are quite viable but usually take several years to bloom. Some mature corms, such as those of *Crocosmia*, live for more than one season and may eventually become very large.

Corms usually have a papery covering that can be peeled away to reveal the fleshy structure beneath. Doing so will also reveal that the upper surface of a corm has growth eyes similar to those of a tuber. Corms may be cut up in the same manner in spring, leaving at least one eye per section.

Bulbs

Tunicate and scaly bulbs

A true bulb does not have growth eyes on its upper surface; instead it produces foliage and roots from a flattened area at the base of the bulb, known as the basal plate. Bulbs come in two types, tunicate and scaly, differentiated by the way their tissue layers, or scales, are arranged. Tunicate bulbs, such as onions and daffodils, have a papery outer skin (the tunica) with scales arranged in concentric rings radiating from the center of the bulb. Scaly bulbs, such as lilies, do not have a papery outer layer, and have smaller overlapping scales, a bit like fish scales, that are usually arranged in a spiral pattern radiating from the center of the bulb.

Tunicate bulbs' protective papery covering and the fact that they form new roots each year allow them to be lifted, dried off completely and stored when dormant. Scaly bulbs cannot withstand dry storage because they have no protective covering and their roots, which live for more than one season, wither if dried.

Bulblets

All bulbs multiply to eventually form clumps, so the obvious method of propagation is to lift and separate the bulbs when they are dormant,

Assorted bulbs and corms: (front left) *Iris* and (front right) *Freesia* with (rear and center) *Watsonia*.

80

growing them on as individual plants. Small bulbs may take a couple of years to flower.

Some bulbs, particularly certain lilies, form clusters of small bulblets or bulbils where the stalk emerges from the bulb. These can be removed and grown on, although they take some time to flower.

Most lilies can be encouraged to form bulblets in the leaf axils along their stems by removing the flower buds and the lower leaves, then mounding up soil around the lower part of the stem. Numerous bulblets will form in the leaf axils and the best will be those around the base, under the mounded soil. A more extreme method is to disbud the stem and wrench it from the bulb, leaving the bulb in the ground. For the best results there should be a few roots at the base of the stem. Lay the stem in a shallow trench and cover it with fine soil, leaving a little foliage exposed. Lift the stem at the end of the season and, with luck, there will be bulblets along its length.

Narcissus bulbs and bulblets (left), *Gladiolus* corms with cormels (right).

Bulb scaling

Many bulbs can be encouraged to form bulblets by damaging their scales. This is known as scaling and encourages bulblets to form at the point of injury, much like the cormels that form where the old corm separates from the new.

Scaling should be done at the end of the growing season when the bulbs are plump and fleshy. Scaly bulbs are easily divided into individual scales; you can remove a few or strip the whole bulb if you wish. Tunicate bulbs must be cut down through the bulb so that each section includes a piece of the basal plate; eight sections is usually about as far as you should go. Stir the divisions up in some moist starter mix in a plastic bag or wide-mouthed jar but do not over-pack the container. Store it in a warm, dark place and within eight to 12 weeks there will be one or more small bulblets at the basal end of each scale or division. Do not remove the bulblets; instead pot each division with the attached bulblets still in place.

Bulb scooping

The bulb scooping method is really only practical with firm tunicate bulbs, such as hyacinths. Firstly, using a spoon or knife, scoop out the basal plate of the bulb. Make sure that every scale is wounded. Dust the cut surface with a fungicide and place the bulb upside down on a wire rack and store in a warm, dark place. It is important that the bulb is warm (64–70°F [18–21°C]), but it must not dry out, so mist it occasionally. The cut surface will contract over a few days, pulling in the outer edges of the bulb. This is quite normal and not a sign of excessive drying. At this stage, the temperature should be increased to around 77–86°F (25–30°C) and in about eight to 12 weeks small bulblets will form on the cut surface. The bulb can now be potted or planted out, still upside down, with the bulblets at or just below soil level.

Method 1: cut shallow wedges out of the cut base of a bulb.

The cuts will open further as the bulb shrivels and bulblets form.

Method 2: hollow out the base of the bulb.

Bulblets will form on the cut surface.

Once the bulblets have formed, place the bulb upside down in a pot, just covering the bulblets with soil.

Bulb scooping

The bulblets will come into growth in spring while the parent bulb gradually withers beneath them. Once they become dormant in autumn they may be lifted and treated as individuals.

Scoring is a simplified version of scooping. Make two intersecting wedge-shaped cuts of about ⅕–⅖ in (5–10 mm) depth across the basal plate of the parent bulb then proceed exactly as for scooped bulbs.

Suckers of sheep laurel (*Kalmia angustifolia*) lifted with roots attached. These can be potted and grown on.

Table 3: Propagation by division

The following table gives information about the size and type of division to use, when to divide, and the period between dividing cycles.

The size — small, medium or large — is in relation to the size of the overall clump. In other words if large divisions are recommended, then a clump may yield only three or four new clumps, but if small divisions are acceptable then it may be possible to make 20 or 30.

The season refers to the optimum time in a reasonably mild climate, one that experiences winter frosts, but where the soil does not freeze solid for days at a time.

The period to maturity is the time the division takes to develop into a medium-sized clump, and the cycle is the time the divisions take to reach the size at which they will themselves be suitable for dividing.

Plant	Season	Size	Mature	Strike	Cycle
Acaena	Any	Natural layer	6–9 months	Good	Yearly
Acanthus	Winter	Large	1 year	Good	Biennially
Achillea	Winter	Small	1 year	Good	Yearly
Aconitum	Winter	Medium	1–2 years	Good	3–4 years
Acorus	Any	Small/medium	1 year	Good	Yearly
Actaea	Win–Spr	Medium	6–9 months	Good	1–2 years
Adiantum	Win–Spr	Small/medium	1–2 years	Good	2–3 years
Adonis	Aut or late Win	Small clumps	3–6 months	Good	2–4 years
Aethionema	Winter	Natural layer	1–2 years	Good	2–3 years
Agapanthus	Any	Large	2 years	Good	2–3 years
Agastache	Spring	Small clump	2–3 months	Good	Yearly
Agave	Any	Rosette	2–3 years	Good	3–4 years
Ageratina	Winter	Medium	1 year	Good	2–3 years
Agrostis	Any	Small clump	3–6 months	Good	Yearly
Ajuga	Aut–Spr	Small/medium	3–9 months	Good	1–2 years
Alcea	Win–Spr	Large	1 year	Moderate	3–4 years
Alchemilla	Win–Spr	Small clump	3–6 months	Good	1–2 years
Allium – ornamental	Winter	Bulbous	1–2 years	Good	Biennially
Aloe	Any	Rosette	1–2 years	Good	2–3 years
Alopecurus	Win–Spr	Small/medium	1 year	Good	Yearly
Alpinia	Winter	Medium	1–2 years	Good	Biennially
Alstroemeria	Aut/Spr	Medium	1–2 years	Moderate	2–3 years
Alyssum	Win–Spr	Few rosettes	3–6 months	Good	1–2 years
Anacyclus	Aut–Spr	Medium clump	6–9 months	Good	1–2 years
Anagallis	Win–Spr	Small clump	2–4 months	Good	Yearly
Anchusa	Aut–Spr	Medium	1 year	Good	1–2 years
Anemone	Aut–Spr	Medium	1–2 years	Good	2–3 years
Anemonella	Autumn	Small/medium	6–9 months	Moderate	2–3 years
Anigozanthus	Aut–Spr	Medium	1–2 years	Moderate	2–4 years
Anthemis	Any	Few rosettes	3–6 months	Good	Yearly

Plant	Season	Size	Mature	Strike	Cycle
Anthericum	Win–Spr	Medium	6–9 months	Good	Biennially
Aptenia	Any	Natural layer	3–6 months	Good	Yearly
Aquilegia	Win–Spr	Large	6 months	Good	Biennially
Arabis	Any	Natural layer	3–6 months	Good	Yearly
Arctotheca	Spr–Aut	Natural layer	3–6 months	Good	Yearly
Arctotis	Spr–Aut	Natural layer	3–6 months	Good	1–2 years
Arenaria	Any	Natural layer	3–6 months	Good	6 months
Arisarum	Aut–Spr	1 good tuber	3–9 months	Good	1–2 years
Arisaema	Win	Small clump	1 year	Good	1–2 years
Aristea	Spring	Small clump	3–6 months	Good	Yearly
Aristolochia	Aut–Win	Medium piece	1 year	Moderate	2–3 years
Armeria	Any	Small clump	3–6 months	Good	6 months
Armoracia	Aut–Spr	1 firm root	3–6 months	Good	6–9 months
Artemisia	Winter	Small/medium	6–9 months	Good	Biennially
Arthropodium	Win–Spr	Medium clump	1 year	Good	2–3 years
Arum	Summer	Medium tuber	1–2 years	Good	Biennially
Aruncus	Win–Spr	Medium clump	1–2 years	Good	2–3 years
Arundinaria	Any	Small clump	3–6 months	Good	Yearly
Arundo	Win–Spr	Medium clump	1–2 years	Good	2–3 years
Asparagus – ornamental	Win–Spr	Medium clump	1–2 years	Moderate	2–3 years
Asphodeline	Win–Spr	Medium	1–2 years	Moderate	2–3 years
Aspidistra	Any	Medium clump	1–2 years	Good	3–4 years
Asplenium	Win–Spr	Medium clump	1–3 years	Good	3–4 years
Astelia	Aut–Spr	Medium clump	6–12 months	Good	2–3 years
Aster	Win–Spr	Small/medium	1–2 years	Good	2–3 years
Astilbe	Win–Spr	Small/medium	1–2 years	Good	2–3 years
Astrantia	Aut–Spr	Small clump	3–6 months	Good	1–2 years
Athyrium	Win–Spr	Small clump	6–9 months	Good	2–3 years
Aubrieta	Win–Spr	Medium/large	6–9 months	Moderate	Biennially
Aurinia	Win–Spr	Few rosettes	3–6 months	Good	1–2 years
Azolla	Spr–Aut	Any piece	1–3 months	Good	3 monthly
Azorella	Aut–Spr	Small clump	3–6 months	Good	1–2 years
Bambusa	Any	Small clump	3–6 months	Good	Yearly
Baptisia	Winter	Small/medium	6–9 months	Good	1–2 years
Begonia – tuberous	Spring	Small	3–6 months	Good	1–2 years
Belamcanda	Aut–Win	Medium	6–9 months	Good	2–3 years
Bellis	Win–Spr	Small clump	3–6 months	Good	Yearly
Bergenia	Aut–Win	Medium	3–9 months	Good	1–2 years
Beschorneria	Aut–Spr	Few rosettes	6–12 months	Good	2–3 years
Billbergia	Spring	1–2 rosettes	6–9 months	Good	2–3 years
Blandfordia	Winter	Medium/large	1 year	Moderate	3–4 years
Blechnum	Win–Spr	Medium clump	6–9 months	Good	1–2 years
Bletilla	Winter	Medium clump	1 year	Good	2–3 years
Boltonia	Winter	Small/medium	3–6 months	Good	1–2 years
Bomarea	Winter	Medium clump	6–9 months	Moderate	2–3 years
Briza	Any	Small clump	2–3 months	Good	Yearly
Bromeliad	Spr–Aut	1–2 rosettes	6–9 months	Good	2–3 years
Brunnera	Win–Spr	Medium	1 year	Good	1–2 years

Plant	Season	Size	Mature	Strike	Cycle
Bulbinella	Aut–Spr	Small clump	1 year	Good	2–3 years
Bupthalmum	Win–Spr	Small clump	6–9 months	Good	1–2 years
Butomus	Spring	Medium	6–9 months	Good	2–3 years
Calamintha	Aut–Spr	Small clump	3–6 months	Good	Yearly
Calla	Spring	Medium	1 year	Good	2–3 years
Calluna	Aut–Spr	Natural layer	9–12 months	Good	1–2 years
Caltha	Aut–Win	Medium	1 year	Good	2–3 years
Calystegia	Aut–Win	Medium	9–12 months	Good	2–3 years
Campanula	Win–Spr	Small/medium	6–9 months	Good	1–2 years
Canarina	Sum–Aut	1 good tuber	3–9 months	Good	1–2 years
Canna	Win–Spr	Medium	1 year	Moderate	2 3 years
Cardamine	Aut–Spr	Small clump	2–6 months	Good	Yearly
Carex	Any	Small clump	2–6 months	Good	Yearly
Cassiope	Any	Natural layer	6–9 months	Good	1–2 years
Catananche	Win–Spr	Small/medium	6–9 months	Good	1–2 years
Cattleya	Post bloom	4 pseudobulbs	6–18 months	Good	2–3 years
Celmisia	Win–Spr	Large	1 year	Poor	3–4 years
Centaurea	Win–Spr	Small/medium	6 9 months	Good	2–3 years
Cephalaria	Winter	Medium	1 year	Good	2–3 years
Cerastium	Aut–Spr	Small/medium	6–9 months	Good	1–2 years
Chamaemelum	Any	1–2 rosettes	3–6 months	Good	Yearly
Chlorophytum	Spr–Aut	Rosette	1 year	Good	Yearly
Chrysanthemum	Win–Spr	Small/medium	6–9 months	Good	1–2 years
Cimicifuga	Aut–Spr	Small/medium	6–9 months	Good	1–2 years
Claytonia	Spring	Offsets	1 year	Moderate	2–3 years
Clematis – herbaceous	Win–Spr	Medium/large	1–2 years	Moderate	3 4 years
Clivia	Spring	Medium clump	1 year	Good	2–3 years
Colchicum	Spr–Sum	Bulbous	1 year	Good	2–3 years
Convallaria	Aut–Win	Small clump	1 year	Good	2–3 years
Coreopsis	Win–Spr	Small clump	3–6 months	Good	Yearly
Coronilla – ground cover	Spring	Natural layer	3–6 months	Good	Yearly
Cortaderia	Win–Spr	Small clump	1 year	Good	2–3 years
Corydalis	Aut–Spr	Small/medium	1 year	Good	2–3 years
Cosmos – perennial forms	Spring	1 tuber	1 year	Good	2–3 years
Cotula	Win–Spr	Small clump	3–6 months	Good	Yearly
Crinum	Win–Spr	Bulbous	1–2 years	Good	3–4 years
Crocosmia	Win–Spr	Bulbous	1 year	Good	2–3 years
Crocus	Sum–Aut	Bulbous	1 year	Good	3–4 years
Cymbalaria	Any	Natural layer	2–3 months	Good	6–9 months
Cymbidium	Post bloom	Pseudobulb	1–2 years	Good	2–3 years
Cymbopogon	Spr–Sum	Small clump	2–3 months	Good	Yearly
Cynara	Spring	Offset, sucker	6–9 months	Good	1–2 years
Cynodon	Spr–Aut	Small sprig	3–6 months	Good	Yearly
Cyperus	Any	Small clump	3–6 months	Good	1–2 years
Cypripedium	Post bloom	3 pseudobulbs	1–2 years	Moderate	3–4 years
Cyrtanthus	Aut–Win	Bulbous	1 year	Good	3–4 years
Daboecia	Winter	Natural layer	1 year	Good	2–3 years
Dactylorhiza	Spring	Large	1–2 years	Moderate	3–4 years

Plant	Season	Size	Mature	Strike	Cycle
Dahlia	Spring	1 tuber	1 year	Good	1–2 years
Davallia	Spring	Small/medium	6–9 months	Good	1–2 years
Delphinium	Win–Spr	Medium	3–6 months	Good	1–2 years
Dianella	Win–Spr	Small/medium	1 year	Good	2–3 years
Dicentra	Win–Spr	Small/medium	3–6 months	Good	1–2 years
Dichondra	Any	Small plug	3–6 months	Good	Yearly
Dictamnus	Winter	Large	1–2 years	Moderate	3–10 years
Dierama	Aut–Spr	Small clump	1 year	Good	2–3 years
Dietes	Winter	Small clump	6–9 months	Good	1–2 years
Digitalis	Aut–Spr	Medium/large	6–9 months	Good	2–3 years
Dioscorea	Aut–Spr	1 good tuber	6–9 months	Good	Yearly
Dodecatheon	Winter	Medium	1–2 years	Moderate	3–5 years
Doodia	Win–Spr	Small clump	6–12 months	Good	2–3 years
Doronicum	Win–Spr	Small/medium	6–9 months	Good	1–3 years
Draba	Spring	Few rosettes	6–9 months	Moderate	2–3 years
Dracocephalum	Win–Spr	Medium	6–9 months	Good	1–2 years
Dracunculus	Win–Spr	Medium	1–2 years	Good	2–3 years
Echeveria	Any	Rosette	3–6 months	Good	Yearly
Echinacea	Win–Spr	Small clump	3–6 months	Good	1–2 years
Echinops	Aut–Spr	Medium/large	1 year	Moderate	2–3 years
Echinopsis	Any	Offsets	1–2 years	Good	Yearly
Epidendrum	Post bloom	4 pseudobulbs	1–2 years	Moderate	3–4 years
Epilobium	Winter	Small clump	3–9 months	Good	2–3 years
Epimedium	Win–Spr	Small clump	6–9 months	Good	1–2 years
Equisetum	Win–Spr	Medium clump	3–6 months	Good	Yearly
Eranthis	Autumn	Medium clump	3–6 months	Good	Yearly
Eremurus	Aut–Win	Medium clump	1 year	Good	2–3 years
Erica	Aut–Spr	Natural layer	9–18 months	Good	1–2 years
Eriophorum	Spring	Medium	6–9 months	Good	1–2 years
Erodium	Any	Small clump	2–3 months	Good	Yearly
Erythronium	Aut–Win	Medium pieces	2–3 years	Moderate	3–5 years
Eucomis	Winter	Small clump	1 year	Good	2–3 years
Eupatorium – herbaceous	Win–Spr	Medium clump	3–6 months	Good	Biennially
Fallopia	Spring	Rooted pieces	6–9 months	Good	1–2 years
Farfugium	Winter	Medium	1 year	Good	2–3 years
Festuca – ornamental	Any	Small clump	2–3 months	Good	6–9 months
Filipendula	Winter	Medium piece	6–9 months	Good	2–3 years
Fragaria	Any	Natural layer	3–6 months	Good	Yearly
Francoa	Winter	Few rosettes	6–9 months	Good	1–2 years
Freesia	Summer	Bulbous	1–2 years	Good	2–3 years
Fritillaria	Aut–Win	Bulbous	2–4 years	Moderate	3–5 years
Gaillardia	Win–Spr	Small clump	3–6 months	Good	Yearly
Galax	Win–Spr	Medium	6–9 months	Good	1–2 years
Galega	Aut–Spr	Small clump	3–6 months	Good	1–2 years
Galium	Aut–Win	Small clump	6–9 months	Good	1–2 years
Gaultheria	Win–Spr	Natural layer	1–2 years	Good	1–2 years
Gaura	Winter	Small clump	6–9 months	Good	1–2 years
Gazania	Aut–Spr	Medium clump	3–6 months	Good	Yearly

Plant	Season	Size	Mature	Strike	Cycle
Gentiana	Aut–Spr	Medium clump	1–2 years	Moderate	3–4 years
Geranium	Any	Medium clump	3–6 months	Good	1–2 years
Gerbera	Spr–Sum	Basal shoot	3–6 months	Moderate	2–3 years
Geum	Win–Spr	Small clump	3–6 months	Good	Yearly
Glechoma	Any	Natural layer	2–3 months	Good	6–9 months
Globularia	Win–Spr	Natural layer, rhizome	2–3 months	Good	1–2 years
Gloriosa	Spring	1 good tuber	1 year	Moderate	1–2 years
Glycyrrhiza	Aut–Spr	Small clump	3–9 months	Good	1–2 years
Gunnera	Win–Spr	Small clump	2–3 years	Good	3–4 years
Gypsophila	Win–Spr	Medium clump	3–9 months	Good	Biennially
Haemanthus	Autumn	Bulbous	1–2 years	Moderate	3–5 years
Haworthia	Any	Few rosettes	6–9 months	Good	Yearly
Hedera	Any	Natural layer	6–9 months	Good	Yearly
Hedychium	Winter	Small tuber	6–9 months	Good	2–3 years
Helenium	Aut–Spr	Small clump	3–6 months	Good	Yearly
Helianthus	Win–Spr	Small clump	3–6 months	Good	Yearly
Helianthus	Spring	1 good tuber	6–9 months	Good	Yearly
Helichrysum – perennial	Any	Medium clump	3–6 months	Good	1–2 years
Helictotrichon	Spring	Small/medium	6–9 months	Good	Yearly
Heliopsis	Win–Spr	Small clump	3–6 months	Good	Yearly
Helleborus	Autumn	Medium clump	1–2 years	Good	2–3 years
Hemerocallis	Winter	Small clump	1–2 years	Good	2–3 years
Hepatica	Aut–Win	Small clump	1–2 years	Good	2–3 years
Hermodactylus	Aut–Win	Small clump	1–2 years	Good	2–3 years
Herniara	Any	Natural layer	2–3 months	Good	6–9 months
Herpolirion	Spring	Medium clump	6–9 months	Moderate	2–3 years
Heterocentron	Any	Natural layer	6–9 months	Good	Yearly
Heuchera	Aut–Spr	1–2 rosettes	6–9 months	Good	Yearly
Hibiscus – perennial	Winter	Large piece	1–2 years	Moderate	3–4 years
Hippocrepis	Any	Natural layer	2–3 months	Good	6–9 months
Hosta	Win–Spr	Small clump	3–6 months	Good	1–2 years
Humulus	Win–Spr	Medium clump	3–6 months	Good	1–2 years
Hyacinthoides	Sum–Aut	Bulbous	1 year	Good	2–3 years
Hypericum – ground cover	Any	Natural layer	6–9 months	Good	Yearly
Hyssopus	Win–Spr	Small clump	3–6 months	Moderate	1–2 years
Iberis – perennial	Winter	Large piece	6–9 months	Moderate	1–2 years
Incarvillea	Aut–Win	Medium clump	6–9 months	Moderate	1–2 years
Indigofera	Spring	Large clump	1–2 years	Moderate	2–4 years
Inula	Win–Spr	Medium clump	9–12 months	Good	1–2 years
Ipheion	Aut–Spr	Small clump	6–9 months	Good	1–2 years
Ipomoea	Winter	Few tubers	6–9 months	Good	Yearly
Iris	Aut–Win	Medium	1–2 years	Good	2–3 years
Jeffersonia	Win–Spr	Large	1 year	Moderate	2–3 years
Juncus	Win–Spr	Medium	1 year	Good	1–2 years
Kniphofia	Spring	Medium clump	1–2 years	Moderate	2–3 years
Lamium	Any	Natural layer	2–3 months	Good	6–9 months
Lathyrus – perennial	Win–Spr	Small clump	3–4 months	Good	Biennially
Laurentia	Any	Natural layer	3–6 months	Good	Yearly

Plant	Season	Size	Mature	Strike	Cycle
Leonotis	Win–Spr	Medium clump	6–9 months	Good	Biennially
Leontopodium	Win–Spr	Medium clump	3–6 months	Moderate	Biennially
Leonurus	Aut–Spr	Medium clump	3–6 months	Good	1–2 years
Liatris	Winter	Medium clump	1 year	Moderate	2–3 years
Libertia	Any	Small clump	1 year	Good	2–3 years
Ligularia	Winter	Medium	1 year	Good	2–3 years
Limonium	Spring	Few rosettes	1 year	Good	2–3 years
Linaria	Winter	Small clump	6–9 months	Good	Biennially
Linnaea	Winter	Medium clump	6–9 months	Moderate	1–2 years
Linum – perennial	Winter	Small clump	6–9 months	Good	1–2 years
Liriope	Any	Small clump	6–9 months	Good	1–2 years
Littonia	Winter	1 good tuber	1 year	Moderate	Biennially
Lobelia – perennial	Aut–Spr	Small clump	6–9 months	Good	2–3 years
Lotus	Spr–Aut	Natural layer	3–6 months	Good	Yearly
Lupinus	Win–Spr	Small clump	6–9 months	Good	Biennially
Lychnis	Win–Spr	Small clump	1 year	Good	Biennially
Lysichiton	Win–Spr	Large piece	1–2 years	Poor	3–5 years
Lysimachia	Any	Natural layer	3–6 months	Good	Yearly
Lythrum	Winter	Large clump	6–9 months	Moderate	Biennially
Macleaya	Winter	Small clump	6–9 months	Good	1–2 years
Mammillaria	Any	Offsets	1–2 years	Good	Yearly
Marrubium	Aut–Spr	Small clump	2–3 months	Good	6–9 months
Mazus	Any	Natural layer	3–6 months	Good	Yearly
Melissa	Aut–Spr	Small clump	3–6 months	Good	6–9 months
Mentha	Any	Small clump	2–3 months	Good	6 monthly
Mertensia	Aut–Win	Small clump	1 year	Good	2–3 years
Microsorium	Any	Firm rhizome	6–9 months	Good	Yearly
Mimulus – ground cover	Any	Small clump	2–3 months	Good	6–9 months
Mirabilis	Win–Spr	1 good tuber	6–9 months	Good	1–2 years
Monarda	Aut–Spr	Small clump	3–6 months	Good	Biennially
Musa	Spr–Sum	Side-shoots	6–18 months	Good	1–2 years
Muscari	Aut–Win	Small clump	6–9 months	Good	Biennially
Myosotis – perennial	Aut–Spr	Small clump	3–6 months	Good	1–2 years
Myrrhis	Winter	Small clump	3–6 months	Good	1–2 years
Nelumbo	Spring	Firm tuber	3–9 months	Good	2–3 years
Neomarica	Win–Spr	Large clump	1 year	Moderate	2–3 years
Nepeta	Any	Small clump	2–3 months	Good	6–9 months
Nephrolepis	Any	Medium clump	3–6 months	Good	Biennially
Nicotiana	Spring	Medium clump	3–6 months	Moderate	1–2 years
Nierembergia	Spring	Small clump	2–4 months	Good	Yearly
Nymphaea	Spring	Medium piece	6–9 months	Good	2–3 years
Odontoglossum	Post bloom	Medium clump	1–3 years	Moderate	3–4 years
Oenothera	Win–Spr	Small clump	3–6 months	Good	Biennially
Omphalodes	Win–Spr	Small clump	3–6 months	Good	1–2 years
Ophiopogon	Any	Small clump	6–9 months	Good	1–2 years
Origanum	Aut–Spr	Small clump	2–3 months	Good	Yearly
Orthrosanthus	Spring	Small clump	3–6 months	Good	Yearly
Ostrowskia	Winter	Root pieces	3–9 months	Good	Biennially

Plant	Season	Size	Mature	Strike	Cycle
Ourisia	Win–Spr	Medium clump	3–9 months	Moderate	2–3 years
Oxalis	Aut–Spr	Small pieces	3–6 months	Good	Yearly
Pachysandra	Aut–Spr	Small clump	6–12 months	Good	1–3 years
Paeonia – herbaceous	Win–Spr	Medium clumps	6–18 months	Moderate	2–4 years
Paeonia – tree	Spr–Aut	Layer, suckers	1–3 years	Moderate	2–4 years
Panax	Win–Spr	Medium clump	2–4 years	Good	3–6 years
Papaver – perennial	Aut–Spr	Medium clump	3–6 months	Good	Biennially
Paphiopedilum	Post bloom	3 pseudobulbs	6–18 months	Good	2–3 years
Parahebe	Any	Clump, layer	3–6 months	Good	Yearly
Pellaea	Aut–Spr	Small clump	3–9 months	Good	1–2 years
Penstemon	Win–Spr	Small clump	3–6 months	Good	Biennially
Persicaria	Spring	Small/medium	6–9 months	Good	Yearly
Philodendron	Spr–Sum	Large clump	6–12 months	Good	2–3 years
Phlomis	Win–Spr	Small clump	3–6 months	Good	1–2 years
Phlox paniculata	Win–Spr	Small clump	3–6 months	Good	Biennially
Phormium	Aut–Spr	Medium clump	6–12 months	Good	2–3 years
Phygelius	Win–Spr	Small clump	2–3 months	Good	Yearly
Phyllostachys	Win–Spr	Rhizome, runner	3–6 months	Good	Yearly
Physalis	Winter	Small clump	3–6 months	Good	1–2 years
Physostegia	Win–Spr	Small clump	3–9 months	Good	Biennially
Phyteuma	Win–Spr	Medium	1 year	Good	2–3 years
Platycodon	Winter	Large pieces	6–9 months	Moderate	2–3 years
Pleione	Win–Spr	Few pseudos	6–12 months	Good	2–3 years
Podophyllum	Aut–Spr	Medium	1 year	Good	1–2 years
Polemonium	Aut–Spr	Small clump	3–6 months	Good	1–2 years
Polianthes	Aut–Win	Side-shoots	6–18 months	Good	Yearly
Polygonatum	Winter	Small pieces	3–9 months	Good	1–2 years
Polygonum	Any	Natural layer	3–6 months	Good	1–2 years
Polypodium	Aut–Spr	Small clump	6–9 months	Good	1–2 years
Polystichum	Aut–Spr	Clump, offsets	6–9 months	Good	1–3 years
Pontederia	Win–Spr	Small clump	3–6 months	Good	Yearly
Potentilla	Any	Runners, layer	6–12 months	Good	1–2 years
Pratia	Any	Clump, layer	2–3 months	Good	6–9 months
Primula – Primrose	Post bloom	Small clump	3–9 months	Good	1–2 years
Primula – Polyanthus	Sum–Aut	Rosette	3–6 months	Good	1–2 years
Pteris	Aut–Spr	Small clump	3–9 months	Good	1–2 years
Pterostylis	Post bloom	3 pseudobulbs	9–18 months	Moderate	2–3 years
Pulmonaria	Aut–Spr	Small clump	3–6 months	Good	Yearly
Pulsatilla	Aut–Spr	Small clump	2–6 months	Good	1–2 years
Ranunculus	Aut–Spr	Small piece or tuber	3–9 months	Good	1–2 years
Raoulia	Spring	Medium pieces	2–3 months	Good	Yearly
Ratibida	Spring	Small/medium	6–9 months	Good	1–2 years
Rebutia	Win–Spr	Rooted stem	3–6 months	Good	1–2 years
Rehmannia	Win–Spr	Small clump	3–6 months	Good	1–2 years
Reinwardtia	Spring	Medium pieces	3–6 months	Good	1–2 years
Rheum	Win–Spr	Medium clump	3–6 months	Good	Yearly
Rhododendron – ev. azalea	Aut–Spr	Natural layer	9–12 months	Good	1–2 years
Rhododendron – low	Aut–Spr	Natural layer	1–2 years	Moderate	2–3 years

Plant	Season	Size	Mature	Strike	Cycle
Rhodohypoxis	Win–Spr	Small clump	3–6 months	Good	2–3 years
Rhus	Aut–Spr	Suckers	1–2 years	Good	1–2 years
Rodgersia	Win–Spr	Small clump	6–12 months	Good	2–3 years
Rohdea	Aut–Spr	Medium clump	3–9 months	Good	1–2 years
Romulea	Aut–Spr	Bulbous	1–2 years	Moderate	3–4 years
Rubia	Win–Spr	Medium clump	6–9 months	Good	Biennially
Rubus – ornamental	Any	Natural layer	3–6 months	Good	Yearly
Rudbeckia – perennial	Win–Spr	Small clump	3–6 months	Good	Yearly
Rumex	Win–Spr	Medium clump	2–3 months	Good	Yearly
Rumohra	Any	Firm rhizome	6–12 months	Good	2–3 years
Ruscus	Aut–Spr	Suckers	6–18 months	Good	2–3 years
Ruta	Win–Spr	Medium clump	2–4 months	Good	Yearly
Sagina	Any	Small clump	2–3 months	Good	6–9 months
Salvia – perennial	Aut–Win	Small clump	3–6 months	Good	1–2 years
Sandersonia	Win–Spr	1 good tuber	6–9 months	Good	1–2 years
Sanguinaria	Aut–Win	Small pieces	3–6 months	Moderate	1–3 years
Sanguisorba	Winter	Medium clump	3–6 months	Moderate	1–3 years
Sansevieria	Spr–Aut	Rosette	6–9 months	Good	2–3 years
Saponaria	Any	Natural layer	2–3 months	Good	6–9 months
Sasa	Any	Small clump	3–6 months	Good	Yearly
Satureja	Aut–Spr	Large clump	2–3 months	Good	1–2 years
Saxifraga	Any	Few rosettes	2–6 months	Good	1–2 years
Scabiosa	Win–Spr	Large clumps	3–6 months	Moderate	2–3 years
Schizostylis	Aut–Spr	Small clump	3–6 months	Good	1–3 years
Scilla	Sum–Aut	Bulbous	3–6 months	Good	1–2 years
Scirpus	Any	Small clump	2–3 months	Good	6–9 months
Scutellaria	Win–Spr	Large clump	6–12 months	Moderate	2–4 years
Sedum	Any	Natural layer	2–6 months	Good	6–9 months
Sempervivum	Any	Rosette	2–6 months	Good	Yearly
Senecio – not shrubby	Winter	Medium clump	2–6 months	Good	1–2 years
Shortia	Winter	Runner, sucker	6–9 months	Good	2–3 years
Sidalcea	Winter	Small clump	3–6 months	Good	1–2 years
Silene	Aut–Spr	Medium clump	3–9 months	Moderate	2–3 years
Sisyrinchium	Any	Small clump	3–6 months	Good	Yearly
Smilacina	Winter	Small pieces	3–9 months	Good	2–3 years
Solanum – tuberous	Win–Spr	1 good tuber	6–9 months	Good	Yearly
Soleirolia	Any	Small clump	1–3 months	Good	3–9 months
Solidago	Win–Spr	Small clumps	3–9 months	Good	1–3 years
Sphaeralcea	Win–Spr	Medium	6–9 months	Good	2–3 years
Stachys	Aut–Spr	Medium clump	3–6 months	Good	2–3 years
Sternbergia	Sum–Aut	Bulbous	6–9 months	Good	2–3 years
Stokesia	Winter	Small clump	3–6 months	Good	1–2 years
Strelitzia	Aut–Spr	Medium clump	1 year	Good	3–4 years
Streptocarpus	Aut–Spr	Few rosettes	3–9 months	Moderate	Biennially
Symphytum	Aut–Spr	Medium clump	2–4 months	Good	1–2 years
Syringa – own roots	Aut–Spr	Suckers	1–2 years	Good	1–2 years
Tanacetum	Aut–Spr	Small clump	2–6 months	Good	Yearly
Tellima	Aut–Spr	Small clump	3–9 months	Good	1–2 years

90

Plant	Season	Size	Mature	Strike	Cycle
Tetrapanax	Spring	Offset, sucker	6–18 months	Good	2–3 years
Teucrium	Win–Spr	Medium clump	3–6 months	Good	2–3 years
Thalictrum	Winter	Medium clump	6–9 months	Good	2–3 years
Thymus	Any	Small clump	2–6 months	Good	Yearly
Tiarella	Win–Spr	Small/medium	6–9 months	Good	1–2 years
Tolmiea	Any	Natural layer	3–9 months	Good	Yearly
Trachelium	Aut–Spr	Medium clump	3–9 months	Good	Biennially
Tradescantia	Any	Small clump	3–9 months	Good	1–2 years
Tragopogon	Win–Spr	Small clump	3–6 months	Good	Yearly
Tricyrtis	Win–Spr	Small clump	6–9 months	Good	Biennially
Trillium	Aut–Win	Medium pieces	3–9 months	Moderate	2–4 years
Trollius	Win–Spr	Small clump	3–6 months	Good	1–3 years
Tropaeolum	Win–Spr	Few tubers	6–9 months	Good	1–3 years
Tulbaghia	Summer	Medium clump	3–6 months	Good	1–2 years
Urceolina	Aut–Win	Bulbous	3–6 months	Good	2–3 years
Uvularia	Win	Medium clump	6–9 months	Good	1–2 years
Vaccinium	Win–Spr	Natural layer	6–9 months	Moderate	2–3 years
Valeriana	Aut–Spr	Small clump	3–6 months	Good	1–2 years
Vancouveria	Aut–Win	Medium clump	3–6 months	Good	2–3 years
Verbascum	Win–Spr	Small clump	3–6 months	Good	1–2 years
Verbena	Any	Clump, layer	2–3 months	Good	6–9 months
Veronica	Aut–Spr	Small clump	3–6 months	Good	Yearly
Viguiera	Winter	Medium clump	3–6 months	Good	1–2 years
Viola	Any	Clump, layer	2–3 months	Good	6–9 months
Wachendorfia	Winter	Few tubers	3–9 months	Good	1–3 years
Wahlenbergia	Aut–Spr	Firm rhizome	3–9 months	Good	Biennially
Xeronema	Win	Large clump	1–2 years	Moderate	3–4 years
Yucca	Spring	Few rosettes	6–9 months	Moderate	2–3 years
Zantedeschia	Win–Spr	Firm rhizome	3–6 months	Good	2–3 years
Zoysia	Any	Small clump	2–3 months	Good	3–6 months
Zygopetalum	Post bloom	Medium clump	6–12 months	Good	2–3 years

6
Propagation by cuttings

Taking a cutting involves removing a piece of tissue, stem, leaf or root from a parent plant and inducing it to form roots or foliage so that it can be grown on as a new plant. Most shrubs and trees and many perennials can be propagated by cuttings.

Woody-stemmed plants that do not grow well from cuttings are usually budded or grafted onto vigorous rootstocks. In many cases the stock for the graft is grown from a cutting, so a knowledge of propagation from cuttings is also important for success in grafting and budding (see Chapter 7).

Stem cuttings

The main types of stem cuttings are herbaceous (or greenwood), softwood, semi-ripe (or semi-hardwood) and hardwood. By selecting the appropriate method it is possible to have cuttings in production year round.

Stem-cutting basics

Cutting stock plants
Most beginners use their garden plants as cutting stock. However, you will often get better results by keeping special stock plants or by preparing your garden plants in advance.

The best cuttings come from vigorous new growth, so prune back intended parent plants a couple of months before taking cuttings to encourage strong new shoots to develop. Of course, you can always use your prunings as cuttings, too.

If you find yourself regularly propagating the same plants, it is often a good idea to pot up a few and keep them as stock plants. That way they can be fed and pruned to produce the maximum growth and if you want cuttings out of season, they can be kept growing indoors.

Root-forming hormones
Plant growth and reproduction are regulated by agents known as phytohormones. They are not true hormones, as found in animals, but act in similar ways.

There are four groups of phytohormones: auxins, gibberellins, cytokinins and inhibitors. The best known is indolebutyric acid, an auxin,

Dahlia cuttings.

Taking a fuchsia softwood tip cutting.

which is found in the compounds often sold as root-forming hormones. Auxins can promote growth or inhibit it. In stems, auxins promote cell elongation and cell differentiation, whereas in roots they inhibit growth in the main roots but promote the development of adventitious roots. Auxins originate at the tips of stems or roots then flow back through the rest of the plant.

Most cuttings are capable of producing enough auxins to stimulate root development. But because it can take quite some time before the roots form we often speed up this process and get better root development by applying the auxin directly at the point where we want roots to form.

Root-forming hormones are available in powder, gel or liquid forms. The powder comes in varying strengths based on the intended use (softwood, semi-ripe or hardwood), while the gel and the liquid can be diluted to any required strength. Knowing what strength to use is a matter of experience as several variables are involved, such as the type of plant, state of growth, time of year and the propagating environment. In most cases, root-forming hormones are not essential, but they tilt the odds of success a little more in favor of the propagator.

Dipping a fuchsia softwood cutting in root-forming hormone powder, which speeds up the process of root development.

Wounding

Cuttings usually produce roots from the cut end at the base of the stem. However, some plants do better if more cambium is exposed. Making a wound on the side of the stem, either as a thin cut or by removing a sliver of bark, may encourage these hard-to-strike plants to develop roots. Some plants, especially rhododendrons, are routinely wounded when taking cuttings, but for most it is something to try if they seem reluctant to strike.

Softwood and semi-ripe cuttings

Many plants, both evergreen and deciduous, can be grown by using cuttings of soft new tip growth. Such softwood cuttings have the advantage of being taken from the most actively growing part of the plant. This means that they strike quickly and develop good root systems, but there are a few disadvantages, too. The young shoots are quite tender and can be damaged when taking the cutting and, unless you have a misting system, they tend to wilt and rot before they strike.

Softwood cuttings are very popular with commercial growers, who usually have sophisticated equipment, but home gardeners may have more success with semi-ripe cuttings. There is no clear-cut dividing line between these two types of cutting: semi-ripe cuttings are just softwood cuttings that have matured a little. In most cases the only difference is the time of year at which the cuttings are taken. Softwood cuttings tend to be most readily available in spring and early summer, while semi-ripe cuttings are more commonly taken in late summer and autumn.

Preparation

Taking a cutting is a fairly standard procedure and does not vary much from plant to plant. The most important factors are timing and the size of the cutting. Although every plant has an optimum time, precise timing is crucial in only a few cases. Evergreen azaleas, for example, can be struck throughout the year, even though the most successful method is to use fairly soft, small, mid-spring to early summer cuttings under mist.

The size of the cutting is usually more important. Beginners often take cuttings that are too large and too mature. Very few cuttings need to be more than three or four nodes long. Softwood cuttings are seldom more than 3 in (7.5 cm) long, while semi-ripe cuttings, with a greater length between the nodes, tend to be larger, typically 4–6 in (10–15 cm) long. Softwood cuttings are, by their very nature, tip cuttings, but semi-ripe cuttings may also be taken from further down the stem. Nevertheless, tip cuttings strike best, regardless of whether they are soft or semi-ripe.

Having taken the cutting off the plant, carefully strip the leaves from the lower nodes, or just the bottom node if the cutting is very small. The softer the cutting, the easier it is to inadvertently strip off the bark as you remove the leaves. Most leaves come away cleanly if they are removed with an upward pulling action after being pulled downwards just enough to break the join between the leaf and the stem.

Once the lower leaves have been removed, trim back the remaining foliage by about half to reduce the transpiration area. Next insert the cuttings in a tray of fresh mix. I prefer a finely sieved 50/50 mixture of bark-based potting mix and perlite. Very soft cuttings may need to be dibbled into place to stop them being bruised, but most cuttings will not be damaged if

Azalea cuttings at various stages of preparation (from left): untrimmed; excess foliage removed; foliage cut back, ready to insert.

Cuttings before and after preparation.

they are gently pushed into place. If the cuttings are going under mist, space them so that they do not overlap. This is important because leaves that are hidden under other foliage will not receive any mist.

If you do not have a misting unit, cover your cutting trays with polyethylene bags or keep them in a humid enclosed propagator. If your cuttings are on a heated pad, always make sure that the bottom heat is not drying out the potting mix, as softwood cuttings usually die if they are allowed to wilt.

Micro-cuttings

Large-leafed plants may be soft and easily damaged, but all the parts are reasonably accessible and easy to get hold of. Not so plants like ericas and callunas with tiny needle-like leaves, or those with very short lengths of new growth, such as many of the small alpine shrubs and perennials.

Assorted micro-cuttings, none more than 1 in (2.5 cm) long. Clockwise from left: *Coleonema*, dwarf *Cryptomeria*, *Santolina*, dwarf juniper and rock phlox.

However, most of these plants are quite easy to propagate. You just have to use very small cuttings. A typical *Erica* or *Calluna* cutting, for example, is no more than about 0.6 in (1.5 cm) long. I call them micro-cuttings and find them some of the most satisfying to work with because it really is a case of creating something from almost nothing.

These plants only rarely strike from anything other than pieces of tip growth. Also, despite appearances, the succulent, tightly foliaged flowering shoots do not, as a rule, strike well. Use the very tip pieces of the non-flowering shoots, which are usually easy to distinguish as they are wirier and often have more widely spaced leaves.

Such tiny cuttings are rather fiddly to work with. They tend to break easily and can be difficult to position in the propagating trays. Use a much more finely sieved mix than you would for larger cuttings and moisten it slightly as you work. The finer soil grains will hold the cuttings better and the moisture will help to bind the soil grains together. Other than those difficulties, micro-cuttings are just like any other tip cuttings, but on a much smaller scale.

Greenwood and basal cuttings

Some shrubby plants, such as *Pelargonium*, do not develop true woody stems, except at the base of very old plants. These sub-shrubs are propagated by a form of softwood cutting known as a greenwood cutting. Personally, I make no distinction between greenwood, softwood and semi-ripe cuttings, but mention the term because you may occasionally see it used.

In spring many herbaceous and tuberous perennials, such as *Begonia*, *Dicentra* and *Delphinium*, produce vigorous, fleshy shoots that emerge directly from their root clumps. These strong-growing basal shoots can be used as softwood cuttings. Avoid letting the shoots get too large, as they strike much faster if taken before the leaves are fully expanded. Most plants that are capable of being grown from basal cuttings develop very quickly once they burst into growth, so there is a limited working time. Taking basal cuttings is a good way to quickly build up stock without having to break up established clumps.

Heel cuttings

Conifer cutting with heel.

A heel cutting is a semi-hardwood cutting that instead of being cut from the parent stock is pulled off so that the cutting comes away with a flap or "heel" of bark at its base. Often used with conifer cuttings, the idea is that the heel firmly anchors the cutting and provides a greater area of exposed cambium. Wherever possible, I prefer to wound cuttings because taking a heel disfigures the parent stock far more than conventional cutting methods.

Leaf-bud cuttings

Climbers often have extremely long internodal length, so traditional stem cuttings tend to be large and unwieldy. It is possible to make smaller cuttings, and get more of them, by using a modified stem cutting known as a leaf-bud cutting. Select a fairly mature leaf and cut the stem just above and below so that you have the leaf and a piece of stem about 2 in (5 cm) long. Trim the leaf back to about half size, dip the base of the stem in root-forming hormone, and insert the cutting.

Honeysuckle leaf-bud cuttings.

If the leaf buds are directly opposite one another, as with *Clematis*, the stem can be split down the middle and each half, complete with leaf and bud, can be used as a cutting.

These techniques can be used only with reasonably firm wood. They work best with evergreen climbers or with deciduous material taken early in the growing season.

Hardwood cuttings

Hardwood cuttings, which are taken in the late autumn and winter when the plants are dormant, are generally used to propagate deciduous shrubs and trees. While many of these plants can be grown from soft or semi-ripe cuttings taken during the growing season, hardwood cuttings are often preferred as they can be left outdoors, which makes them particularly useful if you have no greenhouse or propagating unit.

Hardwood cuttings tend to be quite long, typically 6–12 in (15–30 cm), and are usually struck outdoors in specially prepared beds filled with well-cultivated soil or potting mix. The cuttings are left in the beds over the growing season and lifted in the following winter or early spring.

The long internodes on these honeysuckle stems make them well suited to leaf-bud cuttings.

While outdoor beds are traditional, there is no reason why a cold frame or other propagating unit cannot be used if one is available. Many hardwood cuttings can be struck indoors with bottom heat. Take the cuttings between mid- and late winter and place them on a heated bed or in a frost-free greenhouse. This will induce early growth and they should develop roots as they sprout—several months earlier than they would outdoors.

Mallet cuttings are hardwood cuttings with a small portion of the main stem and a side-shoot. They are most often used if the side-shoots are slender and wiry and therefore likely to move, dry out or collapse during the propagation period. Using a section of the main shoot either side of the side-shoot anchors the cutting as well as increasing the root-forming area. This method can also be used for semi-hardwood cuttings of wiry-stemmed plants, particularly climbers.

Plants with pithy stems, such as hydrangeas, make suitable hardwood cuttings, but the fibrous pith may dry up or rot before the cutting strikes. To get around this problem, cut immediately below a node (where the stem

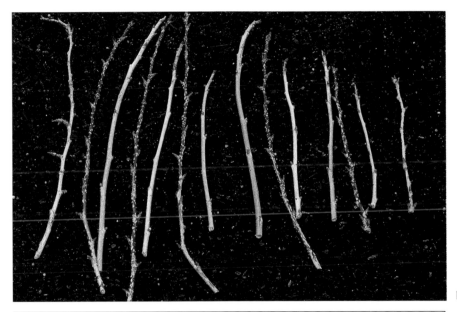

Hardwood cuttings prior to final trimming.

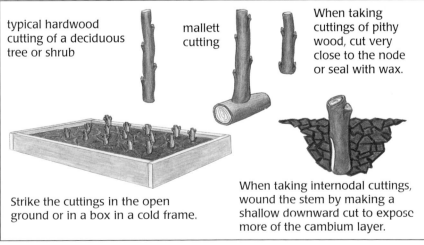

typical hardwood cutting of a deciduous tree or shrub

mallett cutting

When taking cuttings of pithy wood, cut very close to the node or seal with wax.

Strike the cuttings in the open ground or in a box in a cold frame.

When taking internodal cuttings, wound the stem by making a shallow downward cut to expose more of the cambium layer.

Hardwood cuttings.

is at its hardest) and seal the bottom of the cutting with pruning paste or wax. Wound the sides of the stem to expose some of the cambium then insert the cutting in the mix.

Conifers can be struck from hardwood cuttings, although they may take several months to form roots. Hardwood conifer cuttings have foliage, so while it is possible to strike them in the open ground, they tend to dry out and die when exposed to sun and wind. You will get better results using a cold frame or by enclosing the propagating area in a plastic tent, preferably situated in the shade.

Leaf cuttings

Leaf cuttings are an alternative for plants that are not easily divided or that do not produce stems suitable for use as cuttings. They seem to be most successful with plants from tropical or very humid areas.

Leaf cuttings.

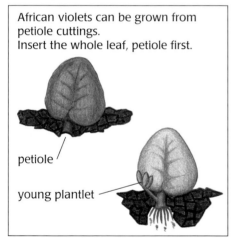

African violets can be grown from petiole cuttings.
Insert the whole leaf, petiole first.

petiole

young plantlet

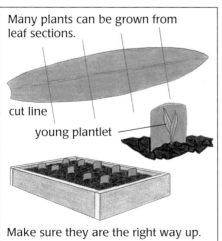

Many plants can be grown from leaf sections.

cut line

young plantlet

Make sure they are the right way up.

Leaf-stem cuttings

There are several different types of leaf cuttings. The most straightforward is the petiole or leaf-stem cutting commonly used with African violets (*Saintpaulia*) and other Gesneriads. Simply remove a leaf from the plant, trim the leaf stem (petiole) to about 1–2 in (2.5–5 cm) long and make a small hole in the potting mix for the base of the leaf. Firm the leaf into place so that it will remain standing up.

When the tray or pot is complete, cover it with a tent of clear plastic film to maintain the humidity and store it in a bright, warm place out of direct sunlight. Bottom heat is a definite advantage and leaf cuttings do well in small heated propagating units.

Leaf sections

With some plants, particularly begonias, cape primrose (*Strepto-carpus*) and *Sansevieria*, sections of the leaves can be used as cuttings. Easiest with long, narrow leaves, the usual method is to cut straight across the leaf and insert the cut end into the soil mix, tip end up. It is also possible to cut the mid-rib from the center of the leaf and use the two halves as cuttings, inserting them cut-side down into the mix.

Leaf slashing

The large-leafed begonias are usually propagated by a slightly different method known as leaf slashing. Remove a healthy leaf and trim the petiole back to about ⅖ in (1 cm) long. Next make cuts through several of the prominent veins on the underside of the leaf, making sure that you go right through the leaf. Using small pieces of wire, pin the leaf, topside up, to a tray of moist cutting mix, and cover the tray with a close-fitting polyethylene cover. After a few weeks, small plantlets will develop on the leaf at the cut points.

Alternatively, cut the leaf into small portions and pin those to the mix. Provided each has a section of vein, they will be able to strike.

Succulents and embryo plants

Many succulents can be grown by leaf cuttings, too, but they are better kept uncovered in less humid conditions. However, it is important to keep the soil moist so that the points of contact between the soil and leaves do not dry out.

Some plants, such as piggy-back plant (*Tolmiea menziesii*) and hen and chickens fern (*Asplenium bulbiferum*), develop small plantlets along the edges of the leaves or at the point where the petiole joins the leaf. Any leaves on which embryo plants develop can be removed and pinned to moist potting mix. The young plantlets will quickly develop roots where they touch the soil or as they drop from the parent leaf. Some succulents, such as *Kalanchoe*, have similar tendencies but prefer a somewhat drier propagating environment.

Aftercare

When to transplant

The ideal time to transplant a cutting is when its root mass roughly equals the amount of foliage it is carrying, which ensures that the plant can be self-supporting. It is usually easy enough to determine when a cutting has struck—it starts to make new growth—but deciding when it should be transplanted is a different matter. You do not want to be too early or you may damage the immature roots, but wait and your cuttings may be weakened by being left in the propagating tray too long. Nevertheless, it is usually better to give the cuttings more time, and provided they are given an occasional liquid feeding they can be left in the propagating tray for a considerable time.

Potting

Except with very small cuttings, which benefit from a finer mix, ordinary all-purpose potting mix is a perfectly suitable medium for freshly struck cuttings. When choosing pots, consider the potential growth rate of the

Three stages of dahlia cuttings (from left): not struck; roots just developing; well-developed roots with the first sign of a tuber forming.

Freshly struck cuttings of bay laurel (*Laurus nobilis*) potted and staked ready for training as standards.

cuttings. While vigorous perennials would do well in small pots, something like an alpine rhododendron would be better in a propagating tube. Over-potting wastes storage space while leaving room for weeds to develop.

Cuttings struck under cover can suffer from moisture stress when exposed to sun or wind and must be gradually hardened before going outside. A week or so in a shadehouse or a shaded part of the garden is usually enough to toughen them up, after which they can be treated as adult plants, fed and then potted as their growth rate demands.

Autumn-struck cuttings

Autumn-struck cuttings pose the problem of deciding what to do with the young plants over winter. In mild, near frost-free climates they can be kept outside, but the colder the climate the more likely it is that such young plants will have to be kept indoors. If the weather is not too extreme, you may be able to gradually harden them in a cold frame then move them out.

Azalea cuttings under mist.

Table 4: Propagation by cuttings

The following table details the most common methods of taking cuttings for particular genera. The time that a cutting takes to strike varies with the conditions. Cuttings under mist or fog tend to take considerably less time than those under polyethylene or in frames. Also misted cuttings may be taken in a softer condition and over a longer period. The strike percentage is also heavily influenced by the method used. Generally, the more primitive the conditions, the lower the strike rate.

| Plant | Type of Cutting | | | | | Season | Time | Strike |
| | Stem Cuttings | | | Leaf | Root | | | |
	Soft	Semi	Hard					
Abelia	✓	✓				Spr/Aut	30	90%
Abeliophyllum		✓				Sum–Aut	28–60	60–80%
Abies		✓				Sum–Aut	50–80	30–75%
Acacia		✓				Spr–Aut	25–80	20–75%
Acanthus					✓	Winter	40–80	90%
Acca	✓	✓				Spr–Aut	30–70	40–75%
Acer	✓	✓				Summer	30–70	50–80%
Ackama		✓				Sum–Aut	30–70	75%
Acmena		✓				Sum–Aut	50	60%
Actinidia	✓	✓				Sum–Aut	20–40	60–95%
Adenandra	✓	✓				Spr–Aut	30–50	50–80%
Aethionema	✓					Spr–Aut	14–35	90%
Agapetes	✓	✓				Spr–Aut	30–60	70–90%
Agathosma	✓	✓				Spr–Aut	21–42	70–90%
Ageratina	✓	✓				Spr–Sum	15–30	90%
Ageratum – perennial	✓					Spr–Aut	15–30	95%
Agonis		✓				Summer	40–60	30–60%
Ailanthus					✓	Spr–Aut	30–45	65%
Ajuga	✓					Any	10–20	100%
Akebia	✓	✓				Spr–Sum	20–40	80%
Albizia		✓				Sum–Aut	20–40	75%
Allamanda		✓				Spr–Aut	25–60	60–80%
Aloe – climbing or bush	✓	✓				Spr–Aut	15–40	90%
Alonsoa	✓					Spr–Aut	7–21	80–90%
Aloysia		✓				Sum–Win	30–100	75%
Alseuosmia	✓	✓				Spr–Aut	30–70	80%
Alyssum – perennial	✓					Sum–Aut	15–40	90%
Amelanchier		✓				Sum	35–50	25–60%
Ampelopsis		✓				Sum–Win	30–100	75%
Anchusa					✓	Sum–Win	15–50	80%
Andromeda	✓	✓				Spr–Aut	30–50	60–90%
Androsace	✓	✓				Spr–Aut	20–30	90%

Plant	Soft	Semi	Hard	Leaf	Root	Season	Time	Strike
Anthemis	✓					Spr–Aut	15–30	90%
Antigonon		✓				Spr–Aut	30–50	60–80%
Antirrhinum – perennial	✓					Spr–Aut	14–28	90%
Aptenia	✓					Spr–Aut	14–28	75–90%
Araujia		✓				Sum–Aut	25–50	70–90%
Arbutus		✓				Spr–Aut	40–80	20–50%
Arctostaphylos	✓	✓				Sum–Aut	30–60	50–80%
Arctotis	✓					Spr–Aut	20–40	70–90%
Ardisia	✓	✓				Spr–Aut	20–40	50–75%
Aristolochia		✓			Win	Summer	20–40	70–90%
Aristotelia	✓	✓				Spr–Aut	25–70	60–80%
Aronia		Sum	Win				28–80	60–80%
Artemisia	✓					Spr–Aut	15–40	80–90%
Asclepias		✓				Sum–Aut	20–40	40–80%
Aster	✓					Summer	10–25	90%
Atriplex	✓	✓				Spr–Aut	15–40	90%
Aubrieta	✓					Any	15–40	90%
Aucuba		✓				Sum–Aut	30–60	80%
Aulax		✓				Aut–Spr	30–80	40–75%
Azara		✓				Sum–Aut	40–60	30–75%
Azorella	✓					Spr–Aut	14–40	75–90%
Backhousia		✓				Autumn	30–70	50–75%
Baeckia	✓	✓				Spr–Aut	21–50	70–90%
Banksia	✓	✓				Sum–Aut	30–80	20–80%
Baptisia	✓					Spring	15–30	70–90%
Bauera	✓	✓				Spr–Aut	28–60	60–80%
Bauhinia	✓	✓				Spr–Aut	20–40	75%
Beaufortia	✓	✓				Spr–Aut	25–50	75–90%
Beaumontia		✓				Sum–Aut	25–50	50–75%
Begonia – tuberous	✓			✓		Spr–Sum	15–40	80%
Bellis	✓					Spr–Aut	10–20	95%
Berberidopsis	✓	✓				Spr–Aut	25–60	60–80%
Berberis		✓				Sum–Aut	40–100	60%
Berzelia	✓	✓				Spr–Aut	21–42	90%
Bignonia	✓	✓				Spr–Aut	25–50	60–80%
Billardiera	✓	✓				Spr–Aut	25–70	30–75%
Boltonia	✓					Spr–Sum	15–40	70–90%
Boronia	✓	✓				Spr–Aut	30–75	50–80%
Bougainvillea	✓	✓				Sum–Aut	30–60	30–75%
Bouvardia	✓	✓			Win	Spr–Aut	25–40	75%
Brachycome	✓					Spr–Aut	10–25	90%
Brachyglottis		✓				Sum–Win	25–80	80%
Brachysema	✓	✓				Sum–Aut	20–40	30–60%
Brugmansia	✓	✓				Sum–Aut	20–40	50–80%
Brunfelsia	✓	✓				Spr–Aut	20–50	75%
Buddleja	Sum	Aut					35–100	80%
Bupthalmum	✓					Spr–Aut	10–28	90%
Buxus	✓	✓				Sum–Aut	35–70	40–80%

Plant	Soft	Semi	Hard	Leaf	Root	Season	Time	Strike
Caesalpinia		✓				Summer	30–50	40–75%
Calamintha	✓					Spr–Aut	7–21	100%
Calceolaria – perennial	✓					Spr–Aut	10–30	90%
Calliandra	✓	✓				Sum–Aut	25–50	60–80%
Callicarpa	Sum	Aut	Win				20–100	40–80%
Callistemon	✓	✓				Sum–Aut	25–40	90%
Calluna	✓					Autumn	25–45	90%
Calocedrus		✓				Sum–Aut	30–80	50–75%
Calocephalus	✓	✓				Sum–Aut	15–40	80%
Calothamnus		✓				Sum–Aut	30–60	30–60%
Calystegia		✓				Sum	20–40	75–90%
Calytrix	✓	✓				Spr–Aut	25–40	80%
Camellia		✓				Summer	40–80	50–80%
Campanula	✓					Spr–Aut	15–30	90%
Campsis	Sum		Win				30–100	80–50%
Cantua	✓	✓				Sum–Aut	20–50	75%
Cardamine	✓			✓		Spr–Aut	10–21	100%
Carica		✓				Spr–Aut	15–40	90%
Carissa	✓	✓				Sum–Aut	25–60	20–60%
Carmichaelia		✓				Spr–Aut	30–70	75%
Carpenteria		✓				Sum–Aut	40–100	10–50%
Caryopteris		Aut	Win				40–100	60%
Cassia	✓	✓				Sum–Aut	25–60	20–60%
Cassinia	✓	✓				Sum–Aut	25–50	40–80%
Cassiope	✓	✓				Spr–Aut	25–50	75–90%
Casuarina		✓				Sum–Aut	30–80	50–75%
Catalpa		✓				Sum	25–50	50–75%
Catananche					✓	Winter	40–70	80%
Cavendishia	✓	✓				Sum–Aut	25–50	40–80%
Ceanothus	✓	✓				Sum–Aut	25–50	75%
Cedronella	✓	✓				Spr–Aut	10–28	90%
Cedrus		✓				Autumn	20–50	30–80
Celastrus					✓	Win–Sp	30–80	75%
Celtis		✓				Sum–Aut	40–60	30–60%
Centaurea – perennial	✓					Spr–Aut	10–30	90%
Ceratopetalum	✓	✓				Sum–Aut	30–60	75%
Ceratostigma	✓	✓				Sum–Aut	25–50	40–75%
Cercidiphyllum	✓	✓				Sum–Aut	30–60	40–75%
Ceropegia	✓					Summer	25–40	75%
Cestrum		✓				Autumn	25–40	75%
Chaenomeles		Sum	Win				25–60	90%
Chamaecyparis		✓				Sum/Win	35–70	40–75%
Chamaemelum	✓					Spr–Aut	10–20	100%
Chamelaucium	✓	✓				Sum–Aut	25–50	30–75%
Choisya	✓	✓				Spr–Aut	25–50	50–85%
Chorizema	✓	✓				Spr–Aut	30–50	50–80%
Chrysanthemum	✓	✓				Spr–Aut	10–20	100%
Cinnamomum		✓				Sum–Aut	25–50	60%

Plant	Soft	Semi	Hard	Leaf	Root	Season	Time	Strike
Cissus		✓				Spr–Aut	25–40	80%
Cistus	✓	✓				Spr–Aut	25–40	75%
Citrus	✓	✓				Sum–Aut	25–50	75%
Clematis		✓				Spr–Aut	25–75	25–80%
Clerodendrum		✓			Win	Spr–Aut	30–50	20–60%
Clethra	✓	✓				Sum–Aut	25–40	50–75%
Clianthus	✓	✓				Spr–Aut	20–40	60–85%
Clytostoma	✓	✓				Spr–Aut	25–50	50–80%
Cobaea	✓	✓				Spr–Aut	10–25	80%
Coleonema	✓	✓				Spr–Aut	25–50	50–85%
Colquhounia	✓	✓				Sum–Aut	20–40	75%
Colutea		✓				Sum–Aut	15–30	85%
Congea		✓				Spr–Aut	25–60	50–75%
Convolvulus	✓	✓				Spr–Aut	14–30	85%
Coprosma	✓	✓				Any	25–50	80%
Corallospartium		✓				Spr–Aut	30–70	75%
Cordyline	✓	✓	✓			Any	21–60+	60–90%
Coreopsis	✓					Spr–Aut	10–20	95%
Cornus	Some	Some	Most				Varies	40–60%
Corokia	✓	✓				Spr–Aut	20–50	40–80%
Coronilla	✓	✓				Spr–Aut	14–30	80%
Correa	✓	✓				Spr–Aut	20–40	75%
Corylopsis	✓	✓				Spr–Sum	20–40	75%
Corylus			✓			Winter	75–100	75%
Corynocarpus	✓	✓				Spr–Aut	28–50	60–80%
Cosmos	✓					Summer	10–25	90%
Cotinus		Sum	Win				30–100	30–75%
Cotoneaster		Sum	Win				30–100	50–80%
Crassula	✓	✓				Spr–Aut	15–30	100%
Crataegus		Sum	Win				40–100	20–50%
Crinodendron	✓	✓				Spr–Aut	30–60	50–75%
Crotalaria	✓	✓				Spr–Aut	15–40	75–90%
Cryptomeria		✓				Any	30–75	75%
Cuphea	✓	✓				Spr–Aut	10–20	100%
Cupressocyparis		✓				Any	40–100	75%
Cupressus		✓				Sum–Aut	30–75	40–75%
Cyananthus	✓	✓				Spr–Sum	15–30	70–90%
Cyathodes	✓	✓				Spr–Aut	30–70	30–60%
Cymbalaria	✓	✓				Any	10–20	100%
Cyphomandra		✓				Spr–Aut	10–30	80%
Cytisus	Sum	Sum	Win				30–80	80%
Daboecia	✓	✓				Spr–Aut	20–40	85%
Dacrydium	✓	✓				Autumn	40–80	10–60%
Dahlia	✓					Spring	10–28	90%
Dampiera	✓	✓				Spr–Aut	20–50	70–90%
Daphne	✓	✓				Spr–Aut	30–75	40–80%
Davidia		✓				Sum–Aut	30–70	30–65%
Delphinium	✓					Spring	10–25	90%

Plant	Soft	Semi	Hard	Leaf	Root	Season	Time	Strike
Desfontainea	✓	✓				Spr–Aut	25–50	40–75%
Deutzia	Sum	Sum	Win				30–100	75%
Dianthus	✓					Spr–Aut	15–30	60–90%
Diascia – perennials	✓					Spr–Aut	10–30	100%
Dicentra	✓					Spr–Sum	10–25	90%
Dictamnus					✓	Spr	30–60	50–75%
Digitalis	✓					Spring	10–25	90%
Dimorphotheca	✓	✓				Spr–Aut	15–30	75–90%
Diosma	✓	✓				Spr–Aut	25–50	60–85%
Diospyros			✓		✓	Spring	30–70	50–75%
Dipelta		✓				Sum–Aut	25–50	75%
Disanthus	✓	✓				Sum–Aut	20–50	50–75%
Distictis	✓	✓				Spr–Aut	15–40	85%
Dodonea		✓				Spr–Aut	30–60	75%
Dombeya		✓				Spr–Aut	15–30	90%
Dracaena		✓				Spr–Aut	30–60	85%
Dracophyllum		✓				Spr–Aut	40–100	20–60%
Dregea		Sum	Win				30–80	50–75%
Drimys	✓	✓				Spr–Aut	20–50	50–75%
Drosanthemum	✓	✓				Spr–Aut	14–28	70–90%
Dryas	✓	✓				Spr–Aut	15–30	80%
Dysoxylum		✓				Spr–Aut	30–80	40–75%
Eccremocarpus	✓	✓				Spr–Aut	15–40	90%
Echinacea	✓					Spr–Aut	10–20	100%
Echinops	Spr				Win		20–50	85%
Echium	✓					Spr–Aut	25–40	50–75%
Edgeworthia	✓	✓				Spr–Aut	25–60	40–80%
Elaeagnus	Sum	Sum	Win				25–75	75%
Elaeocarpus	✓	✓				Sum–Aut	30–50	75%
Embothrium	✓	✓				Spr–Aut	30–60	30–75%
Enkianthus	✓	✓				Sum–Aut	30–70	30–60%
Entelea		✓				Spr–Aut	15–30	90%
Epacris	✓	✓				Spr–Aut	30–50	50–75%
Epilobium	✓					Spr–Aut	15–40	90%
Episcia	✓	✓		✓		Summer	15–25	85%
Erica	✓	✓				Spr–Aut	25–50	85%
Erigeron	✓					Spr–Aut	10–20	100%
Eriobotrya		✓				Sum–Aut	30–70	40–75%
Eriocephalus	✓					Spr–Aut	15–30	90%
Eriostemon	✓	✓				Any	30–70	60–80%
Erodium	✓					Spr–Aut	15–30	80–90%
Eryngium					✓	Win/Spr	30–60	75%
Erysimum	✓	✓				Spr–Aut	15–40	80%
Erythrina	✓	✓				Spr–Aut	30–70	40–80%
Escallonia	Sum	Sum	Win				25–70	80%
Eucryphia	✓	✓				Spr–Aut	30–70	50–75%
Eugenia	✓	✓				Spr–Aut	30–70	40–75%
Euonymus	✓	✓				Spr–Aut	25–60	75%

Plant	Soft	Semi	Hard	Leaf	Root	Season	Time	Strike
Eupatorium – shrubby	✓	✓				Spring	15–30	90%
Euphorbia	✓	✓				Spr–Aut	20–50	75%
Euryops	✓	✓				Any	20–50	85%
Eutaxia	✓	✓				Spr–Aut	30–70	30–60%
Exochorda		✓				Sum–Aut	30–70	40–75%
Fabiana	✓	✓				Spr–Aut	15–35	90%
Fallopia	✓	✓				Summer	20–35	75–90%
Fatshedera		✓				Any	20–40	80%
Fatsia		✓				Any	25–50	80%
Felicia	✓	✓				Any	10–25	100%
Ficus	✓	✓				Spr–Aut	20–60	70–90%
Forsythia	Sum	Sum	Win				25–80	60–90%
Fothergilla	Sum	Sum	Win				30–90	50–75%
Francoa	✓					Spring	20–40	40–75%
Franklinia	✓	✓				Spr–Aut	30–90	30–75%
Fuchsia	✓	✓				Spr–Aut	15–30	95%
Gaillardia	✓					Spr–Aut	10–20	100%
Galphimia	✓	✓				Spr–Aut	21–50	60–80%
Gardenia	✓	✓				Spr–Aut	25–50	60–80%
Garrya		✓				Spr–Aut	40–70	30–65%
Gaultheria	✓	✓				Spr–Aut	25–70	60–80%
Gaura	✓					Spr–Aut	15–30	60–80%
Gazania	✓					Any	10–25	100%
Gelsemium		✓				Spr–Aut	25–60	50–75%
Geniostoma	✓	✓				Spr–Aut	20–60	60–80%
Genista	✓	✓				Spr–Aut	25–50	60–90%
Gentiana	✓					Spr–Aut	20–50	30–75%
Geranium	✓					Spr–Aut	10–30	90%
Geum	✓					Spr–Sum	15–30	90%
Ginkgo	✓	✓				Summer	25–50	50–75%
Glechoma	✓	✓				Any	10–25	100%
Gleditsia			✓			Winter	60–90	50–75%
Goodia	✓	✓				Spr–Aut	25–60	30–75%
Gordonia	✓	✓				Spr–Aut	30–80	25–75%
Grevillea	✓	✓				Spr–Aut	25–60	50–80%
Grewia	✓	✓				Spr–Aut	25–60	50–75%
Greyia	✓	✓				Spr–Aut	25–60	40–75%
Griselinia		✓				Any	25–60	60–90%
Gypsophila	✓					Spr–Aut	20–50	50–80%
Hakea		✓				Sum–Aut	35–50	50–75%
Halimiocistus	✓	✓				Spr–Aut	30–60	75%
Halimium	✓	✓				Spr–Aut	20–50	85%
Hamamelis	✓	✓				Sum–Aut	25–70	20–60%
Hardenbergia	✓	✓				Spr–Aut	20–50	60–90%
Hebe	✓	✓				Any	25–50	85%
Hedera	Sum	Sum	Win				20–50	95%
Hedycarya		✓				Spr–Aut	30–80	50–75%
Helenium	✓					Spr–Sum	15–30	95%

Plant	Soft	Semi	Hard	Leaf	Root	Season	Time	Strike
Helianthemum	✓	✓				Spr–Aut	15–40	60–90%
Helianthus – perennial	✓					Sum–Aut	15–30	90%
Helichrysum – perennial	✓					Spr–Aut	15–30	60–90%
Heliopsis	✓					Spr–Sum	10–25	100%
Heliotropium	✓					Spr–Aut	15–30	75–90%
Heterocentron	✓	✓				Any	20–60	60–80%
Hibbertia	✓	✓				Spr–Aut	25–60	60–90%
Hibiscus – hardy	Spr	Aut	Win				25–80	75%
Hibiscus – tropical	✓	✓				Spr–Aut	20–60	75%
Hippophae		✓				Summer	28–42	50–80%
Hoheria		✓				Sum–Aut	30–70	50–75%
Holmskioldia	✓	✓				Spr–Aut	15–40	90%
Hovea	✓	✓				Spr–Aut	20–60	30–75%
Hovenia	✓	✓				Sum–Aut	30–70	50–75%
Hoya	✓	✓				Spr–Aut	25–60	40–75%
Humulus	✓	✓				Spr–Aut	15–40	75%
Hydrangea	Sum	Sum	Win				15–50	80%
Hymenosporum	✓	✓				Spr–Aut	30–70	40–75%
Hypericum	✓	✓				Any	20–50	80%
Hypoestes	✓	✓				Spr–Aut	14–28	70–90%
Hyssopus	✓					Spr–Aut	10–25	90%
Iberis – perennial	✓	✓				Spr–Aut	20–45	70–90%
Idesia		✓				Sum–Aut	30–50	50–75%
Ilex	✓	✓				Spr–Aut	30–80	50–80%
Impatiens	✓					Spr–Aut	10–20	100%
Indigofera			✓		✓	Win/Spr	40–75	50–75%
Iochroma	✓	✓				Spr–Aut	15–40	70–90%
Ipomoea	✓	✓				Spr–Aut	10–30	70–95%
Isoplexis	✓	✓				Spr–Aut	25–60	70–90%
Itea		✓				Summer	28–42	50–80%
Jacaranda		✓				Sum–Aut	20–50	60–75%
Jasminum		✓				Spr–Aut	20–50	60–80%
Jovellana	✓	✓				Spr–Aut	20–40	85%
Juglans			✓			Winter	60–100	50–75%
Juniperus		✓				Any	30–80	40–75%
Justicia	✓	✓				Spr–Aut	15–35	75–90%
Kadsura	✓	✓				Spr–Aut	20–50	80%
Kalanchoe	✓	✓				Spr–Aut	15–35	90%
Kalmia – not *K. latifolia*	✓	✓				Spr–Aut	30–70	30–75%
Kalmiopsis	✓	✓				Spr–Aut	30–80	50–75%
Kerria	Sum	Sum	Win				25–75	80%
Knautia	✓					Spr–Sum	15–30	75–90%
Koelreuteria					✓	Winter	60–90	75%
Kolkwitzia		Sum	Win				35–90	80%
Kunzea	✓	✓				Spr–Aut	25–60	80%
Laburnum			✓			Winter	60–100	90%
Lagerstroemia	Sum	Sum	Win				30–100	75%
Lagunaria	✓	✓				Spr–Aut	30–70	50–75%

Plant	Soft	Semi	Hard	Leaf	Root	Season	Time	Strike
Lambertia	✓	✓				Sum–Aut	30–70	20–60%
Lamium	✓	✓				Any	10–30	90%
Lampranthus	✓	✓				Spr–Aut	15–40	90%
Lantana	✓	✓				Spr–Aut	15–40	90%
Lathyrus – perennial		✓				Sum	15–35	70–90%
Laurelia	✓	✓				Spr–Aut	30–80	60–80%
Laurus		✓				Sum–Aut	30–70	50–75%
Lavandula	✓	✓				Spr–Aut	25–70	50–90%
Lavatera – perennial	✓	✓				Spr–Aut	15–30	90%
Ledum	✓	✓				Spr–Sum	21–42	60–80%
Leonotis		✓				Spr–Aut	15–40	90%
Leptospermum	✓	✓				Spr–Aut	25–60	80%
Leschenaultia	✓	✓			Win	Spr–Aut	25–70	20–75%
Lespedeza	✓					Spring	15–40	85%
Leucadendron		✓				Sum–Aut	30–90	30–75%
Leucopogon	✓	✓				Spr–Aut	25–90	30–75%
Leucospermum		✓				Sum–Aut	30–90	30–75%
Leucothoe	✓	✓				Spr–Aut	25–70	75%
Levisticum	✓					Spr–Aut	10–30	90%
Liatris	✓					Spr–Aut	15–30	75%
Libocedrus		✓				Spr–Aut	30–100	40–75%
Ligustrum	Sum	Sum	Win				20–80	90%
Lindera		✓				Sum–Aut	28–48	50–75%
Linum	Sum				Win		15–50	60–80%
Liquidambar		Sum			Win		60–100	75%
Lithodora	✓	✓				Spr–Aut	15–40	50–80%
Lobelia – perennial	✓					Spr–Aut	10–30	90%
Lomatia		✓				Sum–Aut	30–90	25–60%
Lonicera	Sum	Sum	Win				25–80	80%
Lophomyrtus	✓	✓				Spr–Aut	30–75	75%
Loropetalum	✓	✓				Sum–Aut	30–70	40–75%
Lotus	✓	✓				Spr–Aut	15–40	60–90%
Luculia		✓				Autumn	30–80	20–60%
Lupinus	✓	✓				Spr–Aut	15–30	90%
Lyonia	Sum	Sum	Win				30–90	50–75%
Lysimachia	✓					Any	10–30	100%
Lythrum	✓					Spr–Aut	15–40	70–90%
Macadamia		✓				Sum–Aut	30–90	20–60%
Macfadyena	✓	✓				Spr–Aut	15–40	75–90%
Macleaya					✓	Win	30–60	50–75%
Magnolia	✓	✓				Spr–Aut	30–90	50–75%
Mahonia		✓				Sum–Aut	30–80	40–80%
Malus – own roots			✓			Winter	60–100	50–80%
Malva	✓	✓				Spr–Aut	14–35	70–90%
Mandevilla		✓				Spr–Aut	20–70	40–80%
Manettia	✓	✓				Spr–Aut	15–40	100%
Marianthus	✓	✓				Spr–Aut	20–50	60–80%
Marrubium	✓	✓				Spr–Aut	15–30	90%

Plant	Soft	Semi	Hard	Leaf	Root	Season	Time	Strike
Maytenus	✓	✓				Spr–Aut	30–70	50–75%
Mazus	✓					Spr–Aut	15–40	70–90%
Melaleuca	✓	✓				Spr–Aut	30–70	40–80%
Melia		Sum			Win		60–100	75%
Melicope	✓	✓				Spr–Aut	25–70	60–80%
Melicytus	✓	✓				Spr–Aut	25–80	60–80%
Melissa	✓					Spr–Aut	10–30	90%
Mentha	✓				✓	Any	10–25	100%
Menziesia	✓	✓				Spr–Sum	21–42	60–80%
Meryta		✓				Sum–Aut	25–60	80%
Mesembryanthemum	✓	✓				Spr–Aut	15–40	75–90%
Metasequoia		Sum	Win				25–60	50–80%
Metrosideros	✓	✓				Spr–Aut	30–80	40–80%
Michaelia	✓	✓				Spr–Aut	25–70	50–75%
Micromyrtus	✓					Spr–Aut	25–60	50–75%
Mimetes		✓				Sum–Aut	30–80	30–60%
Mimulus	✓	✓				Spr–Aut	15–30	90%
Mirbelia	✓	✓				Spr–Aut	25–60	50–75%
Mitraria	✓	✓				Spr–Aut	15–40	90%
Monarda	✓					Spr–Aut	10–30	100%
Monstera		✓				Sum–Aut	20–60	80%
Morus			✓			Winter	75–100	75%
Moschosma	✓	✓				Spr–Aut	15–40	90%
Muehlenbeckia	✓	✓				Spr–Aut	25–60	75%
Murraya	✓	✓				Spr–Aut	25–70	50–75%
Mutisia	✓	✓				Spr–Aut	15–50	70–90%
Myoporum	✓	✓				Spr–Aut	15–40	90%
Myosotis – perennial	✓					Sum–Aut	15–40	90%
Myrsine	✓	✓				Spr–Aut	15–40	90%
Myrtus	✓	✓				Spr–Aut	25–60	60–80%
Nandina	✓	✓				Sum–Aut	25–60	60–85%
Nemesia – perennial	✓					Spr–Aut	10–40	90%
Nepeta	✓	✓				Any	10–30	90%
Nerium		✓				Sum–Aut	35–80	50–75%
Nestegis	✓	✓				Spr–Aut	30–80	50–75%
Nierembergia	✓					Spr–Aut	15–40	90%
Notospartium	✓	✓				Sum–Aut	20–60	30–75%
Nyssa		✓				Sum–Aut	20–50	50–75%
Oenothera	✓				Win	Spr–Aut	15–40	60–80%
Olea	✓	✓				Sum–Aut	40–90	40–75%
Olearia	✓	✓				Spr–Aut	25–70	60–90%
Opuntia		✓				Spr–Aut	25–50	75–90%
Origanum	✓					Spr–Aut	15–30	100%
Osmanthus	Sum	Sum	Win				30–100	50–75%
Osteospermum	✓					Spr–Aut	15–40	90%
Oxydendrum	Sum	Sum	Win				30–100	40–75%
Oxylobium	✓	✓				Spr–Aut	21–50	70–90%
Pachysandra	✓	✓				Any	20–60	85%

Plant	Soft	Semi	Hard	Leaf	Root	Season	Time	Strike
Pachystachys	✓	✓				Spr–Aut	15–40	90%
Pachystegia	✓	✓				Spr–Aut	25–50	60–85%
Pandorea	✓	✓				Spr–Aut	30–70	60–85%
Parahebe	✓	✓				Any	15–30	90%
Paratrophis		✓				Spr–Aut	30–80	50–75%
Parsonsia		✓				Spr–Aut	30–80	40–75%
Parthenocissus		Sum	Win				25–70	90%
Passiflora	✓	✓				Spr–Aut	15–40	90%
Paulownia		✓			Win	Sum	21–42	60–80%
Pelargonium	✓	✓				Spr–Aut	15–40	85%
Pennantia			✓			Winter	60–100	40–70%
Penstemon	✓					Spr–Aut	15–30	90%
Pentas	✓	✓				Spr–Aut	25–50	50–75%
Perovskia	✓	✓				Spr–Sum	20–40	60–80%
Persea		✓				Sum–Aut	30–60	40–75%
Persoonia		✓				Sum–Aut	30–70	25–60%
Petunia	✓					Spr–Aut	10–20	70–90%
Phaenocoma	✓	✓				Spr–Aut	15–35	90%
Phebalium		✓				Sum–Aut	25–60	70–90%
Phellodendron		Sum	Win				28–80	50–75%
Philadelphus		Sum	Win				30–90	60–80%
Philodendron – climbing		✓				Spr–Aut	15–40	70–90%
Phlomis	✓	✓				Spr–Aut	15–30	90%
Phlox – rockery and trailers	✓	✓				Spr–Aut	15–40	70–90%
Phlox paniculata					✓	Aut–Win	40–80	80%
Photinia		✓				Sum–Aut	30–70	60–80%
Phygelius	✓	✓				Any	15–30	90%
Phylica	✓	✓				Sum–Aut	30–70	50–75%
Phyllodoce	✓	✓				Spr–Sum	21–42	60–80%
Physostegia	✓					Spr–Aut	15–30	80–95%
Pieris	✓	✓				Spr–Aut	30–70	60–80%
Pileostegia	✓	✓				Spr–Aut	20–60	60–80%
Pimelea	✓	✓				Spr–Aut	30–60	60–80%
Pisonia		✓				Sum–Aut	25–60	70–90%
Pittosporum	✓	✓				Spr–Aut	30–70	50–80%
Plagianthus		Sum	Win				30–100	40–70%
Platanus			✓			Winter	70–100	80%
Platycodon	✓					Spr–Sum	15–30	85%
Plectranthus	✓	✓				Spr–Aut	15–40	90%
Plumbago	✓	✓				Spr–Aut	25–60	60–90%
Plumeria			✓			Win–Spr	50–100	50–75%
Podalyria		✓				Sum–Aut	40–80	20–60%
Podocarpus	✓	✓				Any	40–100	30–75%
Podranea	✓	✓				Spr–Aut	15–40	90%
Polemonium	✓					Spr–Sum	10–30	100%
Polygala		Any	Win				30–70	50–75%
Polygonum	✓	✓				Spr–Aut	15–40	90%
Pomaderris		✓				Sum–Aut	25–60	60–80%

Plant	Soft	Semi	Hard	Leaf	Root	Season	Time	Strike
Populus			✓			Winter	60–100	90%
Portulacaria	✓	✓				Spr–Aut	15–30	90%
Posoqueria	✓	✓				Spr–Aut	20–50	60–80%
Potentilla	✓	✓				Spr–Aut	25–70	90%
Pratia	✓	✓				Spr–Aut	15–30	90%
Prostanthera	✓	✓				Spr–Aut	30–70	60–80%
Protea		✓				Sum–Aut	40–90	20–75%
Prunus laurocerasus	✓	✓				Spr–Aut	28–60	70–90%
Pseudopanax		✓				Sum–Aut	25–70	60–80%
Pseudowintera	✓	✓				Spr–Aut	30–80	30–65%
Psoralea	✓	✓				Spr–Aut	20–60	70–90%
Pterocarya			✓		✓	Winter	60–100	75%
Pulmonaria	✓					Spr–Aut	15–30	90%
Pulsatilla					✓	Winter	40–80	85%
Punica	Sum	Sum	Win				30–80	75%
Pyracantha	✓	✓				Any	25–60	70–90%
Pyrostegia	✓	✓				Spr–Aut	20–60	60–80%
Quercus		Sum	Win				30–100	20–60%
Quintinia		✓				Spr–Aut	30–80	50–75%
Quisqualis		✓				Sum–Aut	30–70	60–80%
Reinwardtia	✓	✓				Spr–Aut	20–60	70–90%
Rhabdothamnus	✓	✓				Spr–Aut	20–60	80%
Rhamnus		✓				Sum–Aut	30–70	50–75%
Rhaphiolepis	✓	✓				Sum–Aut	30–70	60–80%
Rhododendron – small leaf	✓	✓				Spr–Aut	30–70	60–90%
Rhododendron – dec. azalea	✓	✓				Spring	40–70	30–80%
Rhododendron – ev. azalea	✓	✓				Spr–Aut	30–70	60–95%
Rhododendron – large leaf		✓				Sum–Aut	40–100	30–90%
Rhododendron – vireya	✓	✓				Spr–Aut	40–80	50–90%
Rhodotypos		Sum	Win				28–80	50–75%
Rhus					✓	Winter	40–80	90%
Ribes		Sum	Win				30–100	80%
Robinia – species					✓	Winter	60–100	80%
Romneya					✓	Win–Spr	40–100	30–60%
Rosa – miniature	✓	✓				Spr–Aut	20–50	70–90%
Rosa – own roots	Sum	Sum	Win				30–80	40–90%
Rosmarinus	✓	✓				Spr–Aut	25–60	60–80%
Rubus – Bramble, Raspberry, Blackberry, etc.		Sum	Win				25–80	90%
Rudbeckia	✓					Spr–Aut	10–30	100%
Ruscus	Sum	Sum			Win		40–80	30–60%
Russelia	✓	✓				Spr–Aut	20–60	70–90%
Ruta	✓					Spr–Sum	10–30	95%
Salix			✓			Winter	40–80	100%
Salvia – perennial	✓					Spr–Aut	15–30	90%
Sambucus			✓			Winter	50–80	100%
Sansevieria				✓		Spr–Aut	21–42	90%
Santolina	✓	✓				Spr–Aut	20–40	90%
Sapium			✓			Winter	60–100	75%

Plant	Soft	Semi	Hard	Leaf	Root	Season	Time	Strike
Saponaria	✓					Spr–Aut	15–30	90%
Sarcococca	✓	✓				Spr–Aut	30–70	40–75%
Satureja	✓					Spr–Aut	10–30	90%
Scabiosa	✓					Spr–Aut	20–50	40–80%
Scaevola	✓	✓				Spr–Aut	15–40	90%
Schefflera		✓				Spr–Aut	20–50	90%
Schinus		✓				Spr–Aut	30–70	40–75%
Sedum	✓	✓		✓		Any	10–30	100%
Senecio	✓	✓				Any	20–60	70–90%
Senna	✓	✓				Sum–Aut	25–60	20–60%
Sequoiadendron – dwarf		✓				Any	30–90	50–75%
Serissa	✓	✓				Spr–Aut	15–40	90%
Serruria		✓				Autumn	30–80	20–60%
Shortia	✓	✓				Spr–Aut	30–80	60–80%
Silene	✓	✓				Spr–Aut	15–40	70–90%
Sinningia				✓		Spr–Sum	21–40	70–90%
Skimmia	Sum	Sum	Win				30–90	75%
Solanum	✓	✓				Spr–Aut	15–50	90%
Soleirolia	✓					Any	5–15	100%
Solenostemon (*Coleus*)	✓					Spr–Aut	5–15	100%
Sollya	✓	✓				Spr–Aut	25–60	60–80%
Sophora		✓				Win/Spr	30–100	30–75%
Sorbus		✓				Winter	50–100	25–60%
Sparmannia	✓	✓				Any	25–60	85%
Spartium	Sum	Sum	Win				30–90	60–80%
Sphaeralcea	✓	✓				Spr–Sum	15–30	90%
Spiraea	Sum	Sum	Win				25–90	80%
Stachyurus	✓	✓				Sum–Aut	25–50	75%
Staphylea	Sum	Sum	Win				30–90	80%
Stauntonia	✓	✓				Spr–Aut	40–100	20–60%
Stephanandra			✓			Winter	50–100	80%
Stephanotis	✓	✓				Spr–Aut	35–90	25–60%
Stewartia	✓	✓				Spr–Sum	30–70	25–60%
Stokesia	✓					Spr–Aut	15–30	90%
Streptocarpus				✓		Spr–Aut	20–50	100%
Streptosolen	✓	✓				Spr–Aut	20–60	70–90%
Symphoricarpos			✓			Winter	50–100	80%
Symphytum	✓				Win	Spr–Aut	15–30	100%
Syringa		Sum	Win				30–90	50–80%
Tamarix			✓			Winter	60–100	80%
Tanacetum	✓					Spr–Aut	10–30	100%
Taxodium		✓				Summer	30–50	70–90%
Taxus	Sum	Sum	Win				30–150	40–75%
Tecoma	✓	✓				Spr–Aut	15–40	90%
Tecomanthe		✓				Spr–Aut	25–60	75%
Tecomaria	✓	✓				Spr–Aut	20–60	70–90%
Telopea		✓				Sum–Aut	30–90	20–60%
Ternstroemia	✓	✓				Spr–Aut	30–70	60–80%

Plant	Soft	Semi	Hard	Leaf	Root	Season	Time	Strike
Tetrapanax		✓				Sum–Aut	21–42	90%
Tetrapathea	✓	✓				Spr–Aut	25–60	80%
Teucrium		✓				Any	30–70	60–80%
Thryptomene		✓				Sum–Aut	30–70	60–80%
Thuja		✓				Aut–Win	60–150	75%
Thujopsis		✓				Aut–Win	60–100	80%
Thymus	✓					Any	15–30	100%
Tibouchina	✓	✓				Spr–Aut	25–70	40–75%
Tilia		✓				Autumn	35–80	40–60%
Tolmiea				✓		Any	10–20	100%
Toona					✓	Winter	50–100	75%
Toronia		✓				Sum–Aut	40–100	20–60%
Trachelospermum	✓	✓				Spr–Aut	30–70	40–75%
Tradescantia	✓					Any	10–30	100%
Tristania	✓	✓				Sum–Aut	25–70	40–80%
Tsuga		✓				Sum–Aut	40–100	75%
Tweedia	✓	✓				Spr–Aut	15–40	70–90%
Ulmus			✓			Winter	60–100	75%
Umbellularia		✓				Sum–Aut	30–80	30–70%
Vaccinium		Sum	Spr				30–80	80%
Valeriana	✓					Spr–Aut	15–30	90%
Verbascum	Spr				Win		20–60	90%
Verbena – perennial forms	✓					Any	10–30	100%
Veronica	✓					Spr–Aut	10–30	100%
Vestia	✓	✓				Spr–Aut	15–40	90%
Viburnum		Sum	Win				30–80	60–80%
Viminaria	✓	✓				Spr–Aut	25–60	50–75%
Vinca	✓	✓				Any	10–30	100%
Viola	✓					Any	10–30	100%
Virgilia		✓				Sum–Aut	30–70	40–60%
Vitex		✓				Spr–Aut	30–80	40–75%
Vitis	Sum	Sum	Win				20–80	90%
Weigela	Sum	Sum	Win				25–80	85%
Weinmannia		✓				Spr–Aut	30–80	50–75%
Widdringtonia		✓				Any	30–90	50–75%
Wisteria		Aut	Win				40–100	40–75%
Zelkova					✓	Win	50–100	50–75%
Zenobia	✓	✓				Spr–Aut	30–80	50–75%

7
Less common propagation techniques

Layering

Layering is a natural process most commonly seen in ground covers, which often produce roots at the points where their stems touch the ground. You can simulate this natural layering by keeping a branch in contact with the soil. With time, roots will form and the struck layer may be separated from its parent plant. Most shrubs and trees can be layered, provided they have branches or stems close to the ground.

The method is near foolproof, but it has several drawbacks:

- layers are generally slow to strike (between nine months and a year is typical);
- unless you can stake the growing tip of the layer as it develops you may find that it has lopsided growth;
- layering is limited to fairly small-scale production.

The advantages are:

- layering does not require the removal of material from the parent plant before it has struck;
- the method can be performed *in situ*, with minimal equipment. If you want to propagate a rare plant that provides limited cutting material or a plant that is hard to strike from cuttings and difficult to graft, layering may be the answer.

The process is simple. Choose a branch that can be bent to ground level. Make a shallow cut into the branch, being careful not to seriously weaken it. Dab a little rooting hormone on the wound, peg the stem to the ground with hoops of wire and then mound soil over the wounded area. If it is possible to scratch out a small trench for the stem, so much the better, as mounded soil may blow away.

There are many variations on the layering process that are really just adaptations to suit different plant structures.

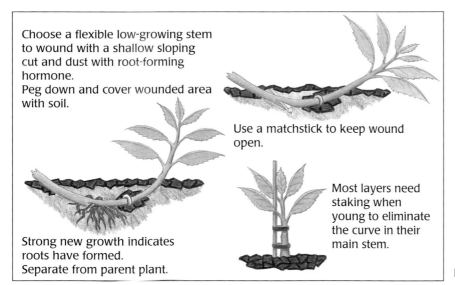

Choose a flexible low-growing stem to wound with a shallow sloping cut and dust with root-forming hormone.
Peg down and cover wounded area with soil.

Use a matchstick to keep wound open.

Most layers need staking when young to eliminate the curve in their main stem.

Strong new growth indicates roots have formed.
Separate from parent plant.

Layering.

Tip layering

Raspberries, blackberries and other plants with long, whippy, cane-like stems strike reasonably quickly if layered at the extreme stem tip. The process is the same as regular layering except that just the tip of the branch is pegged to the soil and buried at the point of contact. Tip layers strike in about six to eight weeks if taken as soon as the canes are long enough to be bent to ground level.

Stooling

Stooling or mound layering works best with deciduous shrubs and is particularly useful when the stems are too inflexible to be bent down to ground level.

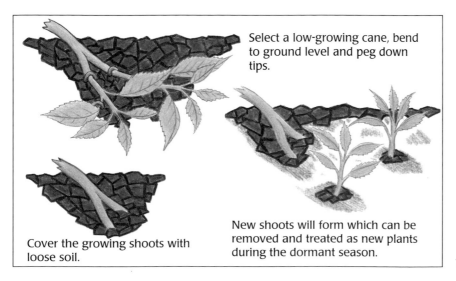

Select a low-growing cane, bend to ground level and peg down tips.

Cover the growing shoots with loose soil.

New shoots will form which can be removed and treated as new plants during the dormant season.

Tip layering.

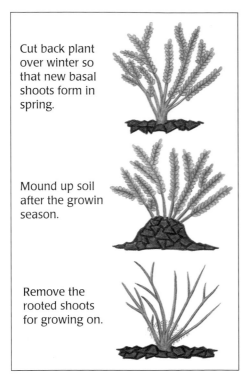

Cut back plant over winter so that new basal shoots form in spring.

Mound up soil after the growin season.

Remove the rooted shoots for growing on.

Stooling.

Cut the parent plant back very hard in late winter—deciduous plants almost to ground level, evergreens not so severely. This hard pruning will cause vigorous basal shoots to develop in spring. As the shoots grow, mound the soil up around them, leaving just the growth tips exposed. Repeat this procedure until late summer, and then allow the stems to grow as normal until winter. When the plant is once again dormant, the soil can be removed and you should find that the stems have developed roots where they have been buried. The newly formed layers can be removed and potted or planted out. The parent plant can be cropped in this way for some years.

Trench layering

Trench or French layering is a variation of stooling. The plants are not hard pruned but very twiggy bushes should be lightly cut back and thinned in winter. About eight to 12 branches some 2 ft (60 cm) long is a manageable number and size. In late winter prepare shallow trenches around the parent plant, bend down the stems and peg them to the ground then mound them over with soil.

As the stems grow, new shoots forming along the branches will push up through the soil. Mound these up with soil, too, leaving just the tips exposed. The following dormant season you can unearth the whole length of the original branch and its side-shoots and divide them up into rooted pieces.

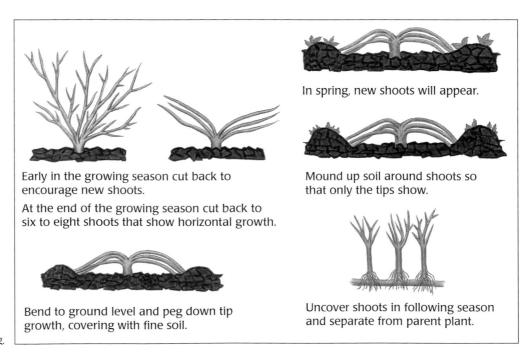

Early in the growing season cut back to encourage new shoots.

At the end of the growing season cut back to six to eight shoots that show horizontal growth.

In spring, new shoots will appear.

Mound up soil around shoots so that only the tips show.

Bend to ground level and peg down tip growth, covering with fine soil.

Uncover shoots in following season and separate from parent plant.

Trench layering.

Dropping

Dropping is similar to stooling, but instead of mounding soil around the plant, the plant is lowered into soil. The method works best with shrubs that develop masses of thin shoots, such as ericas and callunas. Prune the plant, either in winter or early spring, to encourage a large number of shoots. Prepare an area of soil by adding leaf mold, peat or the like so that it is loose and airy. Lift the plant then bury it, angled on its side with just the leaf tips exposed, in the specially prepared soil. By the following winter the plant will have produced numerous struck layers, which you may remove and grow on.

Aerial layering

When you cannot bend a branch down to the soil you have to take the soil to the branch, which is effectively how aerial layering works. It is also a useful technique for plants with large leaves and heavy stems, such as rubber plants (*Ficus elastica*), which are rather unwieldy to handle as cuttings.

Aerial layering does not require any specialized equipment but you will need a sharp knife, some wet sphagnum moss, a patch of black polyethylene sheeting about 1 ft (30 cm) square, a little root-forming hormone (not essential) and some wire ties.

It is best to work with the current year's growth. Choose a reasonably short stem, say about 18 in (45 cm) long, or work near the tip of a branch. The wood should be firm but not completely hardened. Harder wood will strike but it takes considerably longer and the root structure is often poor.

Trim the foliage just above and below the area that you intend to work on so that you have easy access to the stem. Make two slits about halfway around the stem, about ¼ in (5 mm) apart. Remove the strip of bark between the slits, without completely ring-barking the stem. This will stop the wound healing over entirely. If you are working with a fairly soft stem, remove a sliver of bark by making a shallow cut along the stem.

Having made the cut, lightly dust the wound with root-forming powder and wrap the stem with wet sphagnum moss. Hold the sphagnum in position while you secure it with your polyethylene patch and wire ties. It is advisable to have someone help you with this part.

The moss behaves like the soil in conventional layering, preventing the wound drying out and healing over, with the end result that roots form at the wound. The black polyethylene blocks out light, raises the temperature considerably, holds the moss in place and stops it drying out.

Aerial layering is usually done in mid- to late summer, with the plants striking the following spring. However, if you start earlier, say mid-spring, it is possible to have a strike in as little as six to eight weeks. Check the layer by carefully unwrapping the polyethylene. When a good root system has developed, the layer may be removed and grown on.

Aerial layering:
[a] *Magnolia grandiflora* with leaves trimmed and stem wounded.
[b] Stem wrapped in sphagnum moss.
[c] Finished aerial layer. Note green plastic, which is adequate for most uses, though many propagators prefer black because it adds more warmth when exposed to sun.

Grafted rhododendrons sitting on gravel in a container with an inch or so of water. The container is usually covered with glass to create a very humid atmosphere.

Grafting

Grafting combines parts of two plants to make one. It is most commonly used to propagate hybrids and cultivars that fail to strike from cuttings or that produce poor roots.

Grafting is fairly common in nurseries, but few home gardeners practice what is probably the most advanced propagation method that can be done without very sophisticated equipment.

A grafted plant is composed of a stock, which is the part with the roots, and a scion, which is the wood that is to be grafted. All grafting methods require the stock and scion to be cut, trimmed and neatly joined so that their cambium layers match and eventually fuse together. A sharp knife and steady hands are essential. Once joined, the scion is taped in place until it has taken. You will find that successful grafting requires regular practice and is time-consuming, at least at first. It will soon become clear to you why grafted plants are among the most expensive you can buy.

Plants must be reasonably closely related to graft successfully, and even then they may not be completely compatible. Also, there may be problems with varying growth rates between the stock and scion, and with stocks that sucker badly.

The best time for grafting is in spring, as the sap is rising, but if the scion wood starts to grow too early it may collapse before the graft is united. Consequently scion wood is usually collected in late winter, packed so that it remains moist, and kept dormant in cool storage until needed. Although fine with deciduous plants, this is not practical with evergreens. One answer is to pot the stock and keep it indoors over winter so that it comes into spring growth earlier than normal. The scion donor is kept outdoors so that it remains dormant, and the graft is performed once the stock is growing strongly.

Grafting methods

Many grafting methods have been developed, but relatively few are regularly used.

- The **whip and tongue graft** is used when the stock's stem is less than ⅘ in (2 cm) in diameter. It is quite a reliable method but ruins the stock as a useful plant in its own right, so if anything goes wrong all is lost. It also requires that the stock and scion be roughly the same diameter to ensure a close cambium match.
- With **bark grafts**, a large branch is cut and several small scions are inserted under flaps of bark lifted around the cut. This method is most often used in orchards for grafting new varieties onto stocks of older fruit trees. It has the advantages that the scions can be held in place

Robinia pseudoacacia 'Inermis' grafted on to *R. pseudoacacia* stock using a whip and tongue graft.

with small tacks and that the whole graft area is easily sealed with pruning paste.

- **Side grafts** allow the stock be used again if there is an initial failure, and the stock and scion can be different sizes. A drawback is that the grafted branch will always grow at a slight angle.
- **Approach grafting** simulates the natural grafting that may occasionally be seen when two branches remain in contact for a prolonged period. It is usually a last resort, not because it is any less successful than any other means, but because it requires two or more plants to be kept in precise positions until the graft union is well established. Any disturbance before the union is established and the graft is ruined.
- The best methods for home gardeners are the **saddle graft** and its inverted version, the **apical wedge**. These methods require no great carpentry skills and they work well with a wide range of plants. Generally, saddle grafts should be used for evergreens, such as rhododendrons, while the apical wedge works better with deciduous material.

A successful saddle graft on a rhododendron.

Grafts need time to fuse and heal. Be very careful with freshly grafted plants because disturbing the graft union can cause failure. Keep the graft taped until you are sure it has taken and avoid exposing newly grafted plants to moisture stress.

Budding

Budding is commonly used with roses, but is equally successful with many of the wider rose family members, such as apples, quinces, pears and the *Prunus* species.

Budding is straightforward but has one major drawback: a bud is attached to the side of a stem, so there is always a tendency for it to grow outwards before it heads up. Not only does this make the plant hard to shape, it puts weight on the bud union. As this will be a weak point for quite some time, there is a risk of damage, particularly in high winds or if excessive growth is allowed to develop on a newly budded shoot. On the other hand, it is not necessary to destroy the stock plant in order to insert a bud, so if a bud fails to take, nothing is lost.

An apical wedge graft. The graft has not healed completely smoothly.

To produce top-quality plants, you need the best rootstocks. Specialist nurseries grow large quantities of stock plants and can usually be persuaded to part with some, provided you take a decent number. One-year-old stocks are best; plant them out in winter to give them time to get established. However, if you just want to have a go at budding to see how it is done, you can use your existing plants as rootstocks. Try a few roses first.

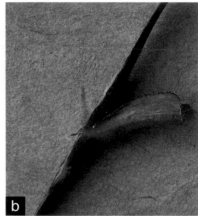

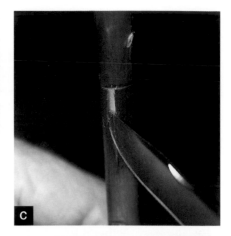

Shield budding:
[a] Removing the bud wood from the bud parent.
[b] Bud ready for insertion. The leaf will drop off naturally and in the meantime can be used as a handle.
[c] Stock with the T-cut made, ready to receive the scion bud.
[d] Scion bud inserted into the T-cut, prior to final trimming and taping.
[e] The bud firmly bound, but with room for some air and light to penetrate.

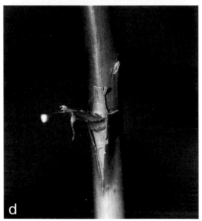

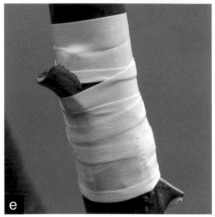

Shield budding

Shield budding, also known as "T" budding, is the technique commonly used with roses. The best time for shield budding is from late spring to late summer when the sap is flowing and the bark is soft and easy to work with, but it can also be done in late winter and early spring.

Start by preparing the rootstock. Work close to the ground, because the closer to the roots the bud is, the less chance there is of stock regrowth. The bark of the stock should be soft, as it must lift easily when cut, but it should not be so soft and green that it starts to wither and dry immediately it is cut. That is why young stocks are best: they still have soft bark near ground level. Trim away any foliage where you intend to bud, then, using a very sharp knife, make a T-shaped cut. The exact size of the cut varies with the size of the bud, typically the stem of the T will be a cut about 1 in (2.5 cm) long, while the cross cut will be about ⅗–⅘ in (1.5–2 cm). Make sure that you do not ring-bark the stem or cut deeply into the wood.

Next, prepare the bud. Look for a mature leaf with a firm bud at the leaf axil. The bud should be plump and fleshy without showing any signs of starting into growth. Remove the leaf, leaving about ⅖–⅗ in (1–1.5 cm) of the leaf stalk. Next, remove the bud by making a shallow cut under it. Start about ⅖ in (1 cm) below the bud, pass under it and continue the cut for about ⅘ in (2 cm) above the bud. This will remove a sliver of bark and stem with bud and the leaf stalk attached. Clean off any parts of woody stem attached to the bark and your bud will be ready to insert.

Gently lift the bark away from the T-shaped cut on the stock, then slide the bud-wood under the bark until the bud is just below the level of the cross cut. Close the flaps of bark back over the bud and trim the surplus bud-wood bark level with the cross cut. Tie the bud using grafting tape, ordinary household tape or plumbers' thread tape, leaving the bud exposed. Rose growers often use specialist budding patches that cover the entire bud.

You will know if the bud has taken within a few weeks. If the stub of petiole drops, leaving the bud looking healthy, then the chances are that the bud has taken. The wound will heal quickly but the bud will not start to grow until the following spring. The tape can be removed and the stem above the bud cut back in late winter or early spring, before growth starts.

In areas of high rainfall, or if the stock plant is inclined to bleed heavily when cut, the incision is sometimes made the other way up, with the top of the T at the base. This will allow rainwater and excess sap to run off, but there is the risk that the bud may fall out or move within the cut.

Chip budding

Chip budding is generally only used when shield budding is not practical, such as with plants that have brittle bark or bark that is difficult to separate from the stem. You may chip bud in summer, autumn or winter and should find that chip buds start to grow sooner than shield buds. However, the failure rate is often higher and the union tends to be weaker.

First, prepare the stock to receive the bud. This requires removing a wedge or chip of bark and stem. Using a very sharp knife, make a cut across and slightly downward into your stock stem. This cut should be at a shallow angle toward the center of the stem but only about ¼ in (5 mm) deep. Next, starting just over 1 in (3 cm) above your first cut, slice downwards along the side of the stem until you intersect the first cut. Neatly remove the wedge of wood and bark, leaving a notch at the base of the cut.

Now, working with the scion, cut an identically shaped chip, with a bud at its center. Insert this chip into the cut on the stock. Make sure that it is locked into the notch at the base of the cut and that the cambium layers match, then tape the chip into position. If the bark on the chips dries and peels then you know the process has failed, otherwise the bud should start to develop in the following growing season.

Chip budding:
[a] stock prepared to receive the scion bud chip.
[b] Scion bud chip taped in place.

Chip budding.

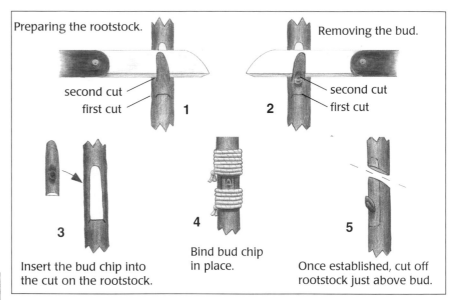

Preparing the rootstock.

Removing the bud.

second cut
first cut
1

second cut
first cut
2

3

Insert the bud chip into
the cut on the rootstock.

4

Bind bud chip
in place.

5

Once established, cut off
rootstock just above bud.

Root cuttings

Some plants may be grown from small lengths of root, though root cuttings are now uncommon because they take a considerable time and often require preparation well in advance of taking the cuttings.

Start by lifting the plant when it is dormant to establish what sort of root structure it has. If there are a number of large fleshy roots that are branching well, then you may take cuttings immediately. If not, you will need to prepare the plant by trimming the roots back to within around 4 in (10 cm) of the crown and then replanting. Strong new roots should develop during the next growing season.

When the roots are ready, lift the plant and remove any excess top growth so that it is easy to handle. Wash the root system, and remove the young fleshy roots near the crown. The original stock may now be replanted. Trim any fine side-shoots away from the roots you have removed so that you have a clean length to work with. The interior of the roots should be white, not brown or discolored. If the roots appear at all diseased, do not use them.

Now trim the roots into pieces of the right length, which is about 4–6 in (10–15 cm) if the cuttings are to be kept outdoors, and 2 in (5 cm) or so for those in heated propagating units.

Use a tray or box filled to about halfway with a good sterile potting or cutting mix. Insert the cuttings vertically without pushing them all of the way in, as they can be covered when you are finished. As there is no easy way of telling which is the top, make sure that you always cut from the top towards the tip, inserting the cuttings as you go. Once the cuttings are all in, fill the box with soil so that the tops of the roots are just level with the

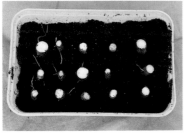

Root cuttings:
[a] Acanthus roots removed, ready for preparation.
[b] Cuttings prepared for insertion in cutting tray.
[c] Cuttings in tray ready for final covering.

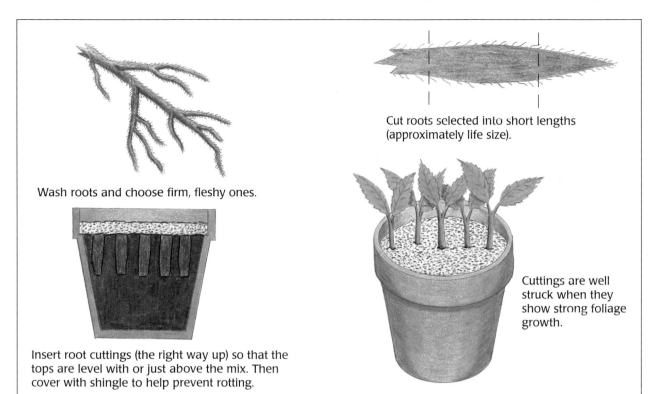

Wash roots and choose firm, fleshy ones.

Cut roots selected into short lengths (approximately life size).

Insert root cuttings (the right way up) so that the tops are level with or just above the mix. Then cover with shingle to help prevent rotting.

Cuttings are well struck when they show strong foliage growth.

Root cuttings.

soil surface. To keep the roots in the dark while preventing them becoming too wet, top up the box with 2/5 in (1 cm) of river sand, perlite or fine pebbles. Do not use root-forming hormones on root cuttings.

You will know the cuttings have struck when leaves start to appear. Allow some good shoots to develop before carefully lifting the cuttings for potting or planting out.

Tissue culture

Tissue culture is the most sophisticated type of propagation. It involves dissecting a growth tip and removing the meristem tissue. All methods rely on meristems for their success, but tissue culture is the only one to use them directly. For this reason it is sometimes known as meristem culture or cell culture.

The meristem is removed under sterile laboratory conditions then induced into growth by the careful use of various phytohormones. For now, tissue culture is restricted to the laboratory and very keen home propagators, but in time it may be able to be used more widely.

PART TWO

TECHNIQUES FOR SOME POPULAR GARDEN ORNAMENTALS

Abelia schumanni

Abelia

Soft or semi-ripe cuttings taken in spring, summer or early autumn usually have a high success rate and most will have roots within 30 days. The cuttings should be semi-ripe green-barked wood because once the bark browns and the wood hardens, the strike rate drops dramatically and the time to strike increases.

Abies fargesii

Abies Fir

Firs may be grown from seed or cuttings. Allow the seed to dry naturally, then stratify for eight to 12 weeks and sow relatively densely at around 59–70°F (15–21°C). Germination is often slow, though reliable. The cultivars must be propagated vegetatively, usually by semi-ripe cuttings of side-shoots taken in summer and early autumn. These take quite a while to strike and the strike rate is sometimes rather poor. Layering also works well and a few forms, such as *A. koreana* 'Compacta,' are grafted.

Acacia baileyana

Acacia Wattle

Wattles are usually grown from seed, which germinates readily at around 68–77°F (20–25°C) and then develops rapidly. Many species have distinctly different juvenile and adult foliage. Semi-ripe cuttings will occasionally strike but are often tricky, and the root development can be poor. Grafts have been tried but not very successfully.

Acanthus mollis

Acanthus Bear's Breeches

Acanthus often self-sows in the garden and collected seed germinates freely. Divide mature clumps every second year in late winter or remove side-shoots with roots and grow on. Vigorous side-shoots without roots can be treated as cuttings.

Acer palmatum

Acer Maple

The species are usually grown from fresh seed, which germinates at relatively cool temperatures of 59–64°F (15–18°C) after stratification. Selected forms are best grafted, commonly using whip and tongue grafts on seedling stock of the same species.

Most maples, especially the vigorous cultivars of *A. negundo*, *A. platanoides* and *A. pseudoplatanus*, will grow from softwood or semi-ripe cuttings kept under mist.

Achillea tomentosa

Achillea Yarrow

Ornamental yarrows are most commonly propagated by division in late winter. Established clumps break up easily and even small divisions quickly start into growth. The species may also be raised from fresh seed, which germinates well if sown uncovered at around 68°F (20°C).

Aconitum Monkshood

These perennials may be divided or raised from seed. Collect the seed as soon as the pods are dry and stratify for several weeks before sowing, lightly covered, with cool temperatures. Even then the germination percentage may not be high. Division in late winter while the clumps are still dormant is easy and reliable provided the divisions are kept on the large side and plants are not divided too regularly—say, every three or four years.

Aconitum napellis

Actaea Baneberry

These tough perennials have graceful foliage and feathery flower heads that are followed by white, red or black berries. Mature clumps divide well in late winter and early spring and the seeds germinate freely at reasonably cool temperatures, around 59–68°F (15–20°C) if sown in spring after stratification.

Aesculus x carnia 'Plantierensis'

Aesculus Horse Chestnut

The various species of these deciduous trees are usually raised from the large seeds, which should be sown covered after a long period of stratification. The young seedlings are very strong so although the seed coat is tough scarification is not really necessary. Hybrids and cultivars must be propagated vegetatively, most often by grafting or budding in the spring. The common horse chestnut (*A. hippocastanum*) is the most readily available rootstock and is easily raised from seed when large numbers are required.

Agapanthus africanus

Agapanthus Lily-of-the-Nile, African Blue Lily

Fresh agapanthus seed germinates very freely at around 72°F (22°C). Plants can also be propagated by dividing established clumps. Do this at any time and in each division include a generous piece of the fleshy, rather bulb-like root.

Ageratina

Closely related to *Eupatorium* and inclined to be rather too vigorous, sometimes invasively so, these large, shrubby perennials may be raised from seed sown in spring or early summer at around 68°F (20°C), from winter or early spring divisions or by taking summer cuttings of non-flowering shoots.

Ageratina ligustrina

Ailanthus Tree of Heaven

If rooted suckers can be removed when the tree is dormant they can be grown on, otherwise take small 2–4 in (5–10 cm) root cuttings in early spring. The seed germinates freely but is not a preferred method because it is not possible to tell the sex of the plant, which is important as male trees have unpleasantly scented flowers that make them less desirable for gardens.

Ailanthus altissima

127

Albizia julibrissin

Albizia Silk Tree

Usually grown from seed, silk trees may also be raised from semi-ripe summer cuttings. Soak the seed for 24 hours and sow it at 74°F (23°C). Selected forms of *A. julibrissin* are sometimes grafted.

Dwarf *Alchemilla* species

Alchemilla Lady's Mantle

Propagation is straightforward by either seed or division. Make divisions in late winter to early spring. Sow the seed as fresh as possible and cover lightly, with cool temperatures, around 59–64°F (15–18°C).

Allamanda schotti

Allamanda

These spectacular shrubs and climbers thrive in tropical and sub-tropical gardens. While they may be raised from seed, the usual propagation method is to take semi-ripe cuttings during the warmer months.

Species with flexible stems, such as the common *A. cathartica*, may also be layered.

Alnus glutinosa (in winter)

Alnus Alder

The seeds, which are often prolifically produced, may be sown lightly covered, either indoors in late winter or outdoors in spring after eight to 12 weeks of stratification. They germinate freely, even in cool conditions. Many species produce suckers that can be removed with roots attached for growing on.

Some selected forms are grafted, usually on to *A. glutinosa* stocks, though with age graft incompatibility often becomes a problem, the stock and scion tending to grow at different rates.

Aloe saponaria

Aloe

Aloes may be propagated by seed, cuttings or, most commonly, by division, which can be done at any time. It is the easiest method but only works for those with root-bearing rosettes at ground level. Remove rooted offsets or easily lifted rosettes and grow them on. Aloes with woody stems can be propagated by removing rosettes and treating them as large cuttings. Sow the seed with heat, around 77°F (25°C), and cover lightly.

Alopecurus pratensis

Alopecurus Foxtail Grass

This genus of grasses with densely tufted flower heads includes both annual and perennial species. The perennials are easily divided from late winter and quickly re-establish to form stocky clumps in summer. They may also be raised from seed, which is the method that must be used for the annual species. The seed can be sown as soon as day temperatures start to routinely reach 64°F (18°C) and it usually germinates within two weeks.

Alstroemeria Peruvian Lily

While selected forms are increased by divisions—usually in spring or in mild climates in autumn—the plants are inclined to take rather a long time to become re-established. Consequently, unless there is a need to perpetuate a particular cultivar, seed is the preferred method. Sow covered, at around 64°F (18°C); it tends to germinate irregularly over several weeks.

Alstroemeria x hybrida 'Sunset'

Amelanchier Juneberry, Serviceberry, Shadbush

These trees and shrubs are often raised from seed and, as might be expected of cool-climate deciduous plants, the seed needs stratifying, usually for around 12 weeks. If your winters are cold enough, simply sow the seed outdoors in autumn and leave it to germinate in spring. Otherwise, stratify and sow indoors in late winter. Summer cuttings strike fairly well and some of the shrubby, suckering types, such as *A. canadensis*, may be divided in late winter.

Amelanchier lamarckii

Andromeda Bog Rosemary

These small, spreading, cool temperate to sub-alpine shrubs—there are only two species—are easily propagated from seed, cuttings or layers, which often occur naturally at points of soil contact. Sow the seed on peaty soil or sphagnum moss in spring, take cuttings in summer and keep them moist or peg down layers at any time. Old, well-established plants with many self-layered pieces can sometimes be divided in spring.

Andromeda polifolia

Anemone Windflower

Bedding anemones are raised from lightly covered seed sown at around 64°F (18°C). The seed of most species can also be treated this way. Rhizomatous and fibrous-rooted species and cultivars can be divided between late autumn and spring depending on the type. Autumn-blooming anemones are best divided in early spring.

Anemone coronaria

Anemonella Rue Anemone

This tiny anemone-like woodland perennial has foliage reminiscent of the maidenhair fern (*Adiantum*) but it is far from delicate. It flowers in spring and early summer and is ready to divide by late summer, when it has become largely dormant. Alternatively, the seed may be sown in spring at around 64°F (18°C).

Anemonella thalictroides

Angelica

Angelica is most often raised from seed, which should be sown as fresh as possible and preferably *in situ*, as the seedlings may not transplant well. Vigorous clumps of perennial types may be divided in spring, though the plants may not react well to this.

Angelica pachycarpa

129

Antirrhinum 'Little Sweetheart'

Arctotis x *hybrida*

Aristolochia grandiflora

Armeria pseudarmeria

Aronia arbutifolia

Artemisia ludoviciana 'Valerie Finnis'

Antirrhinum Snapdragon

Most snapdragons, including the perennial species, are raised from seed sown at around 64–68°F (18–20°C). The seed germinates freely, though the seedlings can be a little reluctant to develop evenly. Perennials, such as *A. hispanicum*, may also be grown from softwood cuttings taken in late spring and summer.

Arctotis African Daisy

Arctotis seeds are most often sown indoors in early spring, lightly covered, at around 59–68°F (15–20°C). Transplant them into individual pots of gritty soil while they are still quite small and plant them out when all frost danger has passed. In suitably mild areas the seed can be scattered on the open ground and left to develop naturally. Well-established clumps often layer naturally and can be broken up. Foliage rosettes separated from the parent plant and treated as cuttings usually strike well.

Aristolochia Dutchman's Pipe, Birthwort

Covering a wide range of growth forms and climate preferences, it is hard to generalize about this genus except to say that all can be raised from seed, which should be soaked before sowing, uncovered, at around 68°F (20°C). Divide herbaceous types in winter while dormant or in very early spring. Semi-ripe stem cuttings are taken in summer and some hardy species may be grown from winter root cuttings.

Armeria Thrift, Sea Pink

Armeria seed germinates well if kept warm and lightly covered and strikes more evenly if soaked for a few hours before sowing. Established clumps are easily broken up at any time but re-establish most rapidly if divided in late winter or early spring.

Aronia Chokeberry

The clusters of striking red or black fruit that are the most appealing feature of these hardy deciduous shrubs contain small seeds that germinate well, after stratification. But because most garden chokeberries are cultivars it is more common to take cuttings, semi-ripe in summer or hardwood in winter. These strike well and in addition to being true to type they will fruit far sooner than seed-raised plants.

Artemisia Wormwood, Mugwort

Artemisias may be herbaceous or shrubby and vary accordingly in their propagation. The shrubby forms grow easily from softwood or semi-ripe cuttings taken in late spring and summer, while the herbaceous types respond well to being divided in late winter or early spring. Seed germinates quickly at moderate temperatures, around 64–68°F (18–20°C).

Arundo Giant Reed

These large rhizomatous perennials die away completely in winter. They can be divided in spring as the new shoots start to appear. Err on the generous side with the divisions and keep them very moist until they are firmly established.

Arundo donax

Asplenium Spleenwort

Most of these ferns have sturdy rhizomatous roots and may be divided, usually in spring. Less vigorous types may be raised from spores and a few species, such as *A. bulbiferum*, develop small plantlets on their fronds that may be grown on as individual plants. The simplest method to establish the plantlets is to remove a mature frond and bury it so that just the small plantlets are exposed. They soon develop roots of their own.

Asplenium oblongifolium

Aster

Asters are grown from seed, division or cuttings, depending on the type. Annuals and species seed may be sown, uncovered, at around 68°F (20°C). Low, clump-forming types, such as *A. alpinus*, can be divided. Those that flower in autumn are divided in spring as the new growth starts, while those that flower in spring and summer are divided after flowering. Non-flowering sideshoots from the main clump can be treated as summer softwood cuttings.

Aster tataricus

Astilbe False Spiraea

Seed of the species is sown at 59–70°F (15–21°C) and may take up to a month to germinate. The species are also often propagated by division. Cultivars and hybrids establish quickly from divisions taken in late winter to early spring as the crowns sprout.

Astilbe cultivar

Astrantia Masterwort

Astrantia seeds germinate quickly if sown lightly covered at around 64°F (18°C). However, as many of the best types are selected forms, vegetative propagation is often necessary, usually by division in spring or autumn. Small, unrooted pieces may be used as cuttings, though mist is often a requirement to keep them turgid long enough for roots to form.

Astrantia major

Aubrieta Aubrietia, Rock Cress

Although often treated as an annual, the garden *A. x cultorum* hybrids are really evergreen perennials. As such, they can be raised from divisions or cuttings as well as by sowing seed uncovered at around 68°F (20°C). Small side-shoots can be used as softwood cuttings at any time and established clumps often yield small rooted pieces, usually in late winter.

Aubretia deltoides

Aucuba japonica

Aucuba Japanese Laurel

This lush-leafed evergreen shrub grows easily from seed sown in the open ground, but the foliage color is often poor and the sex remains unknown until the seedlings mature. Grow selected forms from semi-ripe cuttings at any time. Mist and bottom heat will increase the strike rate and speed up the process.

Azara microphylla

Azara Vanilla Tree

Although *A. microphylla* is the best known, other species also have plenty to offer. Most will grow readily from semi-ripe cuttings taken in summer to autumn. The young plants can be a little tender and often do better if kept in a greenhouse for their first winter. The seed is sometimes hard to obtain but will germinate well if sown lightly covered at around 64°F (18°C).

Babiana stricta

Babiana Baboon Flower

Stopping *Babiana* from propagating itself is more of a problem than growing new plants. These southern African corms produce many small cormlets that can be grown on to flowering size. The corms are usually too small to respond to techniques such as scooping. Seed sown in autumn will flower in around 18 months.

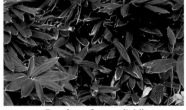

Bamboo *Sasa veitchii*

Bambusa Bamboo

Bamboos may be clump-forming or running (suckering). Both types are easily propagated by division at any time. Other bamboo genera, such as *Pleioblastus*, are usually just as easily propagated. Very large bamboos are difficult to divide because of their sheer size; look instead for small rooted offsets around the base of the clump.

Banksia marginata

Banksia

The species may be grown from seed sown at around 73°F (23°C). Clean the seeds as much as possible and transplant the seedlings while still quite young to avoid root disturbance. Semi-ripe cuttings vary in strike rate; some like *B. integrifolia* are quick and easy while others are difficult. Grow in a low-phosphate potting mix.

Bauhinia galpinii

Bauhinia Orchid Tree

These sub-tropical and tropical trees are easily propagated by seed or cuttings or by removing ready-rooted basal suckers. As many of the best garden forms are cultivars, cuttings are the usual method. Semi-ripe tip cuttings taken from late spring to autumn strike well, especially with mist and bottom heat.

Begonia

Begonias are grown from seed, divisions or cuttings, depending on the type. Bedding begonias and fancy-flowered *B. + tuberhybrida* forms are usually raised from seed. The seed is very fine—dust-like—and is sown uncovered at around 77°F (25°C). Divide tuberous begonias in spring as the new growth starts. Evergreen begonias, such as the beefsteak and rex types, are usually propagated by leaf cuttings, which strike easily but need the constant warmth and humidity of an enclosed propagation unit.

Begonia acutifolia

Bellis Daisy

Bellis daisies are perennials that are usually treated as annuals and raised afresh each year from seed sown uncovered at around 64°F (18°C). Selected cultivars, such as 'Dresden China,' can be grown from late winter divisions or by using unrooted rosettes as softwood cuttings in spring and summer. Watch out for white rust in humid conditions.

Bellis perennis

Berberidopsis Coral Vine

Coral vine is a beautiful and unusual climber that will grow from semi-ripe summer cuttings that are kept cool and moist but it can be unreliable. The best way to secure new plants, although slow, is to peg down layers in autumn. They should have roots by the end of the following summer. Seed sown in spring, lightly covered, at around 64°F (18°C) germinates quite well.

Berberis darwinii

Berberis Barberry

Grow the species from seed and use cuttings or layers for both the species and the selected forms. Evergreen types may be grown from softwood or semi-ripe cuttings; grow deciduous types from semi-ripe cuttings in late summer or from mallet-type hardwood cuttings in winter. Seed of the deciduous Asian species usually requires stratifying.

Bergenia cordifolia

Bergenia Pigsqueak, Elephant's Ear

Bergenia species may be raised from seed sown in spring, uncovered, at around 64°F (18°C). Hybrids and cultivars are best propagated by removing rooted offsets or by dividing well-established clumps in autumn or early winter. Old clumps are often very woody and may be difficult to divide.

Betula Birch

As every gardener knows, birches are very liberal with their seeds and they are not hard to germinate. If you intend to do it deliberately, sow fresh seed outdoors in autumn for spring germination or sow stratified seed in spring. Selected forms, such as the striking white-barked cultivars of *B. utilis*, are usually grafted on to seedling stocks of the same species. Many birches will also grow from summer softwood cuttings under mist, but the strike rate is variable.

Betula pendula (autumn)

133

Bletilla striata

Boltonia asteroides

Boronia molloyae

Bougainvillea 'Carmencita'

Bouvardia laevis

Brachyscome iberidifolia

Bletilla Chinese Ground Orchid

These hardy orchids often multiply quickly in gardens and by far the easiest way to propagate them is simply to divide established clumps in winter. They may take a little while to return to regular flowering but generally their growth continues unchecked.

Boltonia False Chamomile

Popular as cut flowers and in perennial borders, these large daisies are easily divided in late winter when dormant. They may also be raised from spring- or summer-sown seed, which usually germinates within a couple of weeks. Small summer cuttings of non-flowering shoots will strike but need to be well established by late autumn if they are to survive winter dormancy.

Boronia

The species may be grown from seed, but the most common method for species, hybrids and cultivars is to take softwood or semi-ripe cuttings in late summer to autumn. These vary in strike rate but are usually quite reliable and produce good flowering plants for the following spring.

Bougainvillea

Seed sown at 75°F (24°C) germinates well but may take a month or more to show. Cuttings are best taken in summer to early autumn using material that is just starting to harden. The strike rate is often poor and the cuttings are inclined to rot. Occasionally you will get great success, but often it is a struggle. Keep the young plants under cover over winter.

Bouvardia

Mainly grown as short-lived plants for quick color in greenhouses and mild-climate gardens, these perennials and shrubs are usually raised fresh each year from small softwood or semi-ripe cuttings taken in spring. Root cuttings taken in late winter or early spring will also strike but are more difficult to take and offer no real advantages.

Brachyscome Swan River Daisy

These Australian annuals and perennials provide masses of quick color for pots or borders. The annuals are raised from seed sown, uncovered, either indoors or out at around 68°F (20°C). Perennials may also be propagated by taking small tip cuttings of non-flowering shoots during the warmer months.

Brachyglottis

Now including several species formerly classified under *Senecio*, this genus of largely shrubby daisies with a few perennials is most often propagated by taking semi-ripe cuttings. This can be done at any time but is generally most successful in late summer and autumn.

The perennial types are rarely cultivated but may be raised from seed sown, lightly covered, in spring or by breaking up large clumps in late winter or early spring.

Brachyglottis greyii

Bromeliads

Most bromeliads are propagated by carefully breaking up the foliage rosettes. Xerophytes, such as many of the *Tillandsia* species, will grow from any small piece broken off the main plant. If seed is available, sow it in warm, humid conditions, around 77°F (25°C), and keep the young seedlings protected over winter.

Bromeliad *Vriesia carinata*

Brugmansia Angel's Trumpet

Semi-ripe cuttings of these impressive shrubs strike well if taken in late summer. The large leaves need to be cut back to prevent them wilting and to make the cuttings less unwieldy. Mist and bottom heat are definitely advantageous.

Brunfelsia Yesterday, Today and Tomorrow

The several commonly cultivated species of this tropical American genus of shrubs and small trees are most often propagated by taking 2–3 in (5–7.5 cm) long tip cuttings during the warmer months. The cuttings strike quickly if kept in a warm, humid environment.

Brugmansia sanguinea

Buddleja

The seed of most species germinates freely after about 25 days if sown in spring at around 72°F (22°C). The seedlings may even flower in their first year.

Semi-ripe cuttings of the cultivars, especially those of *B. davidii*, strike rapidly and grow quickly.

Brunfelsia australis

Butomus Flowering Rush

Not a rush at all but a lily relative, the sole species in this genus thrives in the very damp ground around pond or stream margins. Its seeds germinate well if sown at around 68°F (20°C) but can be difficult to keep at just the right level of dampness.

Divisions are usually easier and established clumps break up well in early spring.

Buddleja 'Lochinch'

135

Buxus sempervirens

Buxus Box

Box is a plant that is often required in large numbers for hedging, so it helps to be able to propagate your own. While the species will grow from spring-sown seed, the usual requirement is for plants of a known size and foliage type, so cuttings are the normal propagation method. Use softwood or semi-ripe cuttings taken in late spring to early autumn and be prepared for a variable strike rate that can be frustratingly slow.

Caesalpinia pulcherrima

Caesalpinia

While these beautiful sub-tropical and tropical shrubs and small trees will strike readily from summer semi-ripe cuttings and layers, they are often raised from seed. Soak the seed well, for at least 24 hours, and sow it, covered, at warm temperatures. The seeds may also be pre-sprouted on damp paper towels before sowing.

Herbeohybrida *Calceolaria*

Calceolaria Slipper Flower

The bedding or pot calceolarias are raised from seed sown uncovered at around 68°F (20°C). The sowing time varies depending upon whether you want a warm season garden display or indoor flowers in the cooler months. The shrubby perennial types, such as *C. integrifolia*, are easily raised from spring to autumn softwood cuttings.

Calla Water Arum, Bog Arum

The true calla lily, not to be confused with *Zantedeschia*, which is often called calla, is a bog perennial from the northern cool-temperate zone. It is most easily propagated by division in very early spring and while sometimes slow to re-establish it is usually reliable.

Callicarpa pedunculata

Callicarpa Beautyberry

These shrubs or small trees vary in hardiness, some being quite frost tender. Cuttings can be taken through most of the year, softwood or semi-hardwood in the warmer months and hardwood from late autumn through winter. The colorful drupes for which most species are grown contain small seeds that should be sown at around 68°F (20°C) with a light soil covering. The seed of hardy species may need stratification.

Callistemon 'Mauve Mist'

Callistemon Bottlebrush

Bottlebrush seed germinates very freely over a wide temperature range but may take a considerable time to flower. Semi-ripe summer to autumn cuttings are the way to propagate hybrids and cultivars. They usually strike well and in most cases will flower the following season.

Camassia Quamash

These American bulbous perennials quickly form large clumps that can be divided in late winter or early spring. The seed grows readily too and should be sown in light soil, covered, at around 64°F (18°C) as soon as it is ripe.

Camassia quamash

Camellia

If you are raising species or your own hybrid crosses, camellia seed should be soaked before sowing at 68–75°F (20–24°C). Germination often takes up to two months. Cuttings of the many hybrids and cultivars are best taken in late spring to early summer as the new growth is starting to firm up. The strike rate is variable and may take some time. Mist and bottom heat help.

Camellia 'Aubrey Magill'

Campanula Bellflower

Campanulas may be annual, biennial or perennial, trailing, clump-forming or rather bushy, and their propagation varies accordingly. The annuals and biennials are raised from lightly covered seed sown at around 68°F (20°C). Perennial trailers are propagated by removing small rooted pieces or by using lengths of stem as cuttings. Divide clump-forming types in winter, early spring or after flowering, but the taller, bushier types are usually best divided in late winter.

Campanula x pulloides

Campsis Trumpet Vine

These vigorous deciduous climbers are known for producing masses of suckers. These, along with winter root cuttings, are the simplest way to propagate a few new plants. Spring softwood cuttings will strike under mist and summer semi-ripe cuttings can be left in a cool greenhouse or a shaded frost-free place outdoors to strike over winter. Spring-sown seed germinates well at around 68°F (20°C) but usually needs a period of stratification.

Campsis grandiflora

Caragana Peashrub

These rather rangy shrubs could seldom be called neat but they do have unusually colored pea-like flowers and interesting, inflated seedpods. While some forms, such as the weeping *Caragana arborescens* 'Pendula' are grafted onto seedling stocks, most plants are raised from seed. The seeds are very hard and must be scarified and/or soaked before being sown, covered, at around 68°F (20°C).

Caragana arborescens

Cardiocrinum Giant Lily

Although these tall bulbous perennials can be divided, seed is by far the most common method of propagation and much quicker for building up good numbers of plants. However, seedlings take several years to reach flowering size. Sow the seed as soon as it is ripe, covered, at around 68°F (20°C). The seedlings will need to be well established before winter if they are to survive dormancy and are best kept sheltered for their first winter.

Cardiocrinum giganteum

137

Carex flagellifera

Carpinus betulinus

Carya illinoinensis

Castanea sativa

Allocasuarina equisetifolium

Catalpa bignonioides

Carex Sedge

Like most sedges, grasses and reeds, the *Carex* species and cultivars are very easily divided at almost any time, so much so that it is rare to raise them from seed. If you wish to try seed, sow it uncovered at around 64°F (18°C). Provided the seed is fresh, it should germinate quickly.

Carpinus Hornbeam

The species are usually raised from stratified seeds sown in spring or fresh seed sown in autumn that can be left to chill naturally over winter. The seed germinates erratically and may take two years. For small numbers, selected forms can be raised from layers of the lower branches, otherwise grafting onto seedling stock is necessary.

Carya Hickory, Pecan

These beautiful deciduous trees with rough-textured bark, bright fall foliage, hard wood and often-edible nuts are surprisingly uncommon in gardens. However, they are not difficult to propagate.

Seed is the usual method and it should be sown outdoors in fall or stratified for 8–12 weeks. It germinates in around three weeks at around 68°F (20°C). Selected forms are usually budded or grafted onto seedling stocks.

Castanea Chestnut

Chestnuts can be sown outdoors as soon as ripe or may be given light winter chilling indoors and sown in spring after scarifying and soaking. Chestnuts should be sown covered at around 64°F (18°C). The best nuts usually come from named clones and these must be grafted onto seedling stocks.

Casuarina She Oak, Horsetail Tree

Casuarina seed is very fine and germinates freely if sown barely covered, in summer, on light, rather sandy soil. The seedlings grow rapidly and should be potted into individual containers as soon as possible to enable them to develop a sturdy root system. Semi-ripe summer and autumn cuttings will root but are rather slow and often have a low strike rate.

Catalpa Indian Bean

This very attractive, showy-flowered, broadleaved, deciduous tree may be raised from seed or cuttings and some forms are better budded in summer or grafted onto seedlings using the whip and tongue method. The seed germinates more evenly if stratified for a few weeks and should be sown, lightly covered, at around 68°F (20°C).

Summer cuttings of non-flowering stems strike quite well but need to be well established before winter.

Ceanothus California Lilac

Ceanothus species will grow easily from seed, but as most of the garden types are hybrids or cultivars they are usually propagated by cuttings. Use softwood or semi-ripe tip cuttings taken from mid-summer to autumn and treat the struck cuttings carefully because their roots are very brittle.

Ceanothus papillosus 'Roweana'

Cedrus Cedar

Cedars produce large quantities of seed and it is not difficult to germinate. Germination rates are better and the time is shorter if the seed is stratified for eight weeks or so. Sow the seed at around 68°F (20°C) and cover it. Cuttings will strike, but they are variable. Try semi-ripe cuttings in autumn. Some selected forms are grafted.

Cedrus sp.

Centaurea Cornflower, Knapweed

Annual cornflowers are raised from late winter- or early spring-sown seed. Sow the seed at around 64°F (18°C) and cover it lightly. Perennial knapweeds may be grown from divisions or cuttings. Divide established clumps in late winter to early spring or take cuttings of the vigorous young spring shoots or from the basal shoots that appear through the growing season.

Centaurea cyanus

Ceratostigma Plumbago

The more herbaceous species, such as *C. plumbaginoides*, may be lifted and divided in spring, while the shrubby types like Chinese plumbago (*C. willmottianum*) are usually propagated from summer cuttings or layers.

The seed, if not sown fresh in autumn in the open, benefits from a brief stratification and germinates at moderate temperatures, around 64°F (18°C).

Ceratostigma plumbaginoides

Cercis Redbud, Judas Tree

Most species produce copious quantities of seed and this is the easiest way to propagate them. Soak the seed in warm water then stratify it for around 12 weeks and sow it, covered, in spring at around 72°F (22°C). The cultivars of the Judas tree (*C. siliquastrum*) and eastern redbud (*C. canadensis*) are usually grown from summer softwood cuttings.

Cercis siliquastrum

Cestrum

Most species are very easily propagated by taking summer to autumn cuttings. The seed germinates quickly if it is soaked before sowing and then kept warm. As they tend to resent root disturbance the seedlings and struck cuttings should be moved into individual pots as soon as possible.

Cestrum nocturnum

139

Chaenomeles 'Chocharagaki'

Chamaecyparis lawsoniana 'Pembury Blue'

Chimonanthus praecox

Chionodoxa luciliae

Choisya ternata

Chrysanthemum 'Bullfinch'

Chaenomeles Flowering Quince, Japonica

Often only too keen to propagate themselves from seed, flowering quinces should be propagated vegetatively to ensure the cultivars remain true to type. Semi-ripe summer cuttings and winter hardwood cuttings strike equally well. It may be also possible to remove rooted suckers or, if only a few new plants are required, try layering. Budding and grafting are options too, though rarely necessary. If you want to sow seed, over-winter it outdoors or stratify before sowing.

Chamaecyparis False Cypress

There are hundreds of *Chamaecyparis* cultivars and unless you intend to grow the species, the usual method of propagation is semi-ripe cuttings taken from summer to early winter. Although the strike rate varies with the cultivar, most present no great difficulties. Mist and bottom heat will increase the strike rate and greatly decrease the striking time.

Chimonanthus Wintersweet

Wintersweet is usually grown from spring-sown seed that has been chilled for 12 weeks. It germinates freely over a wide temperature range but the seedlings take a considerable time to flower, usually five or more years. Grafts and layers have been tried with limited success.

Chionodoxa Glory of the Snow

Glory-of-the-snow bulbs are so small that there is really no way to propagate them other than to let them multiply naturally. Fortunately, they do this readily and large clumps soon develop, allowing the bulbs to be lifted and separated during the summer dormant season. The seed germinates freely if sown in early spring, but takes three years to produce flowering bulbs.

Choisya Mexican Orange Blossom

The species and cultivars strike well from semi-ripe cuttings taken at any time during the growing season, but are especially quick in late summer and early autumn. In mild climates, winter hardwood cuttings will strike outdoors if they are kept shaded and moist.

Chrysanthemum

Florist's chrysanthemums (*Dendranthema* x *grandiflorum*) are hybrids grown from cuttings or divisions. Divisions are usually made in early spring as growth begins, while cuttings can be made at any time from spring until dormancy. Chrysanthemums will germinate freely from seed sown in spring at 70°F (21°C). Marguerite daisies (*Argyranthemum frutescens*) strike very easily from cuttings at any time, as do shastas (*Leucanthemum*). With mist and bottom heat it is possible to produce vigorous chrysanthemum cuttings in as little as two weeks.

Cimicifuga Bugbane

Although fresh seed germinates well, division is the most straightforward and immediate propagation technique. Large clumps broken up at any time from late autumn to early spring soon re-establish themselves.

Cimicifuga racemosa

Cinnamomum Camphor Tree, Cinnamon

Seed is not always available and even then the seedlings can be variable with regard to foliage size and color and bark texture. Consequently, taking semi-ripe, summer to autumn cuttings from superior trees is the preferred propagation method. Mist and bottom heat will greatly improve the strike rate and constant warmth also helps.

Cinnamomum camphora

Cissus Kangaroo Vine

Semi-ripe *Cissus* cuttings strike very well, especially with bottom heat and mist. The young plants grow rapidly and because they may become entangled with each other they should be transplanted into individual pots as soon as possible.

Cistus Rock Rose

Although the species grow well from seed, most garden plants are grown from cuttings. Use soft or semi-ripe wood from late summer to autumn. The strike rate is usually high, especially with mist and bottom heat.

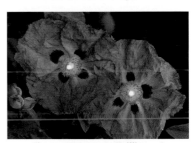
Cistus x purpureus 'Brilliancy'

Citrus Orange, Lemon, Grapefruit, etc.

Although it is fun to raise citrus plants from seed (pips), the main reason for doing so is not to perpetuate the best varieties but to provide stocks on to which superior plants may be budded or grafted. Another plant, *Poncirus trifoliata* (related, but not in the genus *Citrus*) is also often used as a stock.

Citrus seed germinates well at moderate to warm temperatures and if the seeds' hard outer coating can be peeled away the germination will be more even. Seedlings intended for grafting or budding are usually allowed to grow for 18–24 months and are then most often budded using the inverted T method.

Poncirus trifoliata

Plants on their own roots are usually grown from summer to autumn semi-ripe cuttings, often taken with a heel, or by layering. As citrus tend to be woody with few branches near the ground, aerial layering may be necessary for those that are reluctant to grow from cuttings.

Claytonia Spring Beauty

These small western North American alpines sometimes produce offsets that can be removed in spring and grown on. Otherwise, seed is the method of propagation. It should be stratified for a few weeks and then sown, very lightly covered, on gritty soil at around 68°F (20°C). The seedlings are very prone to damping off and should be kept just moist.

Citrus meyeri

Clematis 'Nellie Moser'

Clerodendrum splendens

Clethra arborea

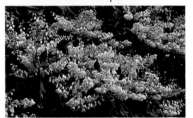

Clianthus puniceus

Clivia miniata

Cobaea scandens

Clematis Virgin's Bower

Most species will germinate from seed in about two months. The seed should be stratified and sown at around 72°F (22°C). Herbaceous species, such as *C. integrifolia* can be divided when dormant. Cuttings of the cultivars and hybrids vary in their strike rate but are often very good. *Clematis montana* forms may be grown from semi-hardwood cuttings in the growing season or hardwood cuttings over winter. *Jackmanii* hybrids are best grown from semi-ripe cuttings from late spring. They vary in their strike rate and are inclined to collapse as young plants.

Clerodendrum Glory Bower, Glory Flower, Butterfly Bush

These shrubs, erect or twining, vary considerably in hardiness and growth habit. While the climbers, such as *C. splendens* and *C. thomsoniae*, are often raised from seed which requires warm temperatures to germinate, most are propagated from semi-ripe summer to autumn cuttings. Late winter to early spring root cuttings, rooted suckers and layering are other options.

Clethra Summersweet, Lily-of-the-Valley Tree

Although most species are easy enough to raise from spring-sown seed, cuttings are generally preferred. Summer to autumn semi-ripe cuttings strike best with a small heel. Seedlings and freshly struck cuttings are prone to collapsing if stressed. Many species sucker freely and rooted suckers are a quick way to obtain a few new plants. Lift the suckers in early spring.

Clianthus Kaka Beak, Parrot's Bill

Clianthus puniceus seed germinates well, especially if soaked before sowing, but propagation by cuttings often produces stronger plants. Cultivars must be vegetatively propagated. Use softwood or semi-ripe cuttings from late spring to mid-autumn. They strike well and quickly come into flower.

Clivia Kaffir Lily

Division is the most common way of propagating *Clivia*. Large clumps are easily broken up in summer when at their most dormant. The divisions may be reluctant to flower in their first season but soon settle down. The seed may be sown at around 73°F (23°C). It should be fresh and lightly covered. Seedlings will not flower until at least three years old.

Cobaea Cup and Saucer Vine

The only commonly cultivated species in this genus is a quick-growing climber. It is usually propagated from seeds, which germinate well even at cool temperatures if sown indoors in late winter or outdoors in spring. Seeds are flat and are best placed on their edges to prevent rotting. Softwood and semi-ripe cuttings taken throughout the warmer months also strike well and should be used to perpetuate the cultivars, such as 'Alba.'

Colchicum Autumn Crocus

These corms form clumps that can be lifted and divided when at their most dormant in late spring and very early summer. If available, the seed can be sown, lightly covered, in a sandy soil. Small cormlets and seedlings may take several years to flower.

Colchicum bornmuelleri

Coleonema

This genus is often confused with the very similar but not very closely related *Diosma*. These genera contain a number of small evergreen shrubs with fine, whippy stems and minute, aromatic leaves. They are easily propagated during the warmer month from cuttings of the non-flowering stems. The roots are quite brittle, so take care when potting the freshly struck cuttings.

Coleonema pulchrum 'Pinkie'

Convallaria Lily-of-the-Valley

Lily-of-the-valley quickly forms large clumps of tangled rhizomatous roots. The clumps can be broken up at any time, but do best if divided in late winter, just as the growth shoots are becoming obvious. The plants may initially resent the disturbance, but usually recover quickly.

Convallaria majalis

Coprosma

Although *Coprosma* seed germinates quickly at around 68°F (20°C), most garden plants are cultivars raised from cuttings. Softwood or semi-ripe cuttings of most species and cultivars strike well at any time, particularly in late summer and autumn.

Coprosma repens

Cordyline Ti Tree, Cabbage Tree

This genus includes several relatively hardy types from New Zealand and two tender tropical species, *C. stricta* and *C. terminalis*, that are often grown as houseplants. Plants are raised either from seed or by using lengths of stems or the stem tips as cuttings. The seed germinates well if sown in spring at around 68°F (20°C) but is really only suitable for propagating species. The fancy cultivars can be propagated at any time by cutting lengths of stem and laying them on their sides, half-buried, in a warm, moist propagating frame. Rooted shoots will grow along the stems. Aerial layering and removal of suckers are also often possible.

Cordyline australis

Coreopsis Tickseed

Annuals are raised from lightly covered seed sown at around 68°F (20°C). The perennial species may also be raised from seed, but it is best left uncovered and slightly cooler—around 64°F (18°C). Hybrids and cultivars are propagated by division or cuttings. Clump-forming types are best divided in late winter or very early spring, while cuttings of strong basal shoots can be taken at any time in the growing season and strike quickly.

Coreopsis verticillata

Cornus florida

Corokia x *virgata*

Coronilla valentina ssp. *glauca* 'Variegata'

Corydalis wilsonii

Corylopsis sinensis var *calvescens*

Cosmos bipinnatus

Cornus Dogwood

Most species will grow freely from seed but will usually require stratification, some, such as *Cornus mas*, for up to four months. Depending on the parent species, selected forms are grown from cuttings, suckers or are grafted: *C. baileyii* and *C. stolonifera* is propagated by suckers; *C. mas* from seed or grafts; *C. nuttallii* from layers or cuttings; and *C. florida* from cuttings or grafts. Consult more specialized information for precise details.

Corokia

These tough evergreen shrubs are often grown for their drupes but are seldom raised from seed as most garden forms are cultivars. Instead semi-ripe cuttings are taken, mainly in summer and autumn, or the lower branches may be layered. If species are required the seed germinates well at around 64°F (18°C) and can be extracted from the fruit by steeping in water.

Coronilla Crown Vetch

Seed sown in spring germinates at around 64°F (18°C) and should be scarified or soaked before sowing to soften the seed coat. The spreading types often self-layer and rooted pieces can be removed in summer. Softwood and semi-ripe cuttings of the shrubby species may be taken through the warmer months. In all cases be careful of the brittle roots and pinch back the young plants to encourage bushiness.

Corydalis

The species may be raised from seed, but division is the most common propagation method for both species and cultivars. Established clumps can be broken up any time from autumn to early spring and will quickly regenerate.

Corylopsis Winter Hazel

These winter-flowering deciduous shrubs and small trees are raised from seed, summer cuttings or layers. Sow the seed outdoors as soon as it is ripe or stratify indoors for sowing in the spring. Semi-ripe cuttings strike well but must be well established by autumn if they are to over-winter outdoors. Layers put down in early spring can usually be lifted late in the following winter.

Cosmos Mexican Aster

Propagate annual bedding by seed: sow covered at around 68°F (20°C). Divide the perennial forms with tuberous roots in early spring, or strike from summer cuttings of the vigorous basal shoots.

Cotinus Smoke Bush

Smoke bushes are usually propagated using fairly large, late-spring to early-summer softwood cuttings. It is important to have the young plants growing well by leaf fall or they will very likely collapse during the winter. Winter hardwood cuttings also strike, though less successfully.

Cotinus coggyria

Cotoneaster

Propagate the evergreens from semi-ripe tip cuttings taken from late summer until late autumn. Start earlier with the deciduous types to give them time enough to harden off before leaf fall. The strike rate is quite variable and depends greatly on the individual species or cultivar. The species may be raised from stratified seed, sown covered at 64–68°F (18–20°C).

Cotoneaster franchetti

Crassula

In common with many succulents, the crassulas are easily propagated from cuttings. Tip pieces strike best, though if you look around the base of may species you may find pieces with aerial roots attached or that have self-layered. Leaf cuttings also strike well and the seed is not difficult to raise on a gritty seed mix.

Crassula rupestris

Crataegus Hawthorn, May

Raise the species from lightly covered stratified seed sown at around 64–68°F (18–20°C), but propagate the cultivars and hybrids by semi-ripe tip cuttings in summer or hardwood cuttings in winter. The strike rate is variable, often quite low. Some forms are grafted.

Crataegus phaenopyrum

Crinodendron

These beautiful South American shrubs prefer cool, moist conditions and are most often propagated from semi-ripe cuttings taken from late spring until autumn. Although mild bottom heat will help, keep the foliage cool to prevent wilting.

Crinodendron hookerianum

Crinum

These very large bulbs form congested clumps that, apart from the sheer effort of lifting and separating them, are not difficult to divide, usually in early spring. Offsets will form around the edge of the clump and require less effort to remove but it is better for the vigor of the plants to divide the whole clump.

Fresh seed germinates quickly at around 68°F (20°C) and the seedlings flower in around three to four years.

Crinum x powellii

145

Crocus versicolor 'Picturata'

Cryptomeria japonica var *sinensis*

Cuphea micropetala

Cupressus macrocarpa

Cyclamen persicum

Crocus

Provided the growing conditions are not too warm, crocuses usually multiply quite quickly and are propagated by simply lifting the multiplied corms when dormant and replanting them later.

Most garden varieties do not set seed, but seed of the wild species is occasionally available. Sow it fresh outdoors in autumn, or stratify and sow in the spring.

Cryptomeria Japanese Cedar

There is only one species of *Cryptomeria*, but many cultivars. These are nearly all cutting-grown and strike quite readily from semi-ripe to near hardwood cuttings taken from late summer until early winter. A few that are difficult to root are grafted.

Cuphea

Sometimes treated as annuals, these small soft-wooded shrubs often self-sow profusely and propagation is then simply a matter of lifting the seedlings and growing them on. Alternatively, harvest the fresh seed and sow, lightly covered, at around 68°F (20°C). Softwood and semi-ripe cuttings strike very quickly.

Cupressus Cypress

These conifers can be propagated by seed, cuttings or by grafting. Seed is the easiest way to get large numbers of species plants for shelter or hedging. Stratify the seed for a few weeks then sow in spring at around 68°F (20°C).

Cultivars and hybrids can be raised from semi-ripe cuttings under mist. These sometimes strike better if taken with a heel. Those that are difficult to strike may have to be grafted on to seedling stock with similar parentage.

Cyananthus

Closely related to the bellflowers (*Campanula*), these Himalayan alpine perennials resent root disturbance and are consequently difficult to divide. The seed germinates well if sown, lightly covered, in spring and most species will also grow from soft and semi-ripe spring and summer cuttings of non-flowering stems.

Cyclamen

Apart from a few of the more vigorous small-flowered species, *Cyclamen* tubers are slow to multiply and hard to divide. Consequently, most are raised from seed, which should be as fresh as possible and not dry. Seed sown lightly covered at around 64°F (18°C) germinates quickly and the seedlings may flower in their first season.

Cydonia Quince

Once popular for producing a uniquely flavored jelly, the best fruiting forms of quince are grafted onto seedling stocks, which can be raised by taking the seed from ripe fruit, cleaning off the fruit pulp, stratifying for 8–12 weeks, then sowing in spring. At 68°F (20°C) the first seedlings should appear within three weeks and will be suitable for use as stocks from their second year onward.

Cydonia oblonga

Cymbidium

These widely grown orchids occur in a myriad of cultivars and are popular with home hybridists. Named forms are propagated vegetatively by dividing established clumps of the pseudobulbs, which break up easily. Division should be done early in the growing season and often stimulates a flush of growth. Even the leafless backbulbs will sprout leaves, though the flowers may be poor in the first season.

Seedlings are usually raised on a jelly medium in glass flasks— something for the real enthusiast.

Cymbidium Fifi 'Henry' x Insigne 'Album'

Cynara Cardoon, Globe Artichoke

Resembling large thistles, these large, quick-growing perennials are usually propagated by dividing established clumps in late winter. If you are reluctant to break up such large plants, try removing some of the offsets that often form around the base. They may come away with roots attached or they can be treated as cutting.

Lightly covered seed sown at around 68°F (20°C) germinates well but will take a few years to reach maturity.

Cynara carduncularis

Cyperus Umbrella Plant, Papyrus

Large clumps are easily divided in early spring and the seed develops quickly if sown in damp soil at around 68°F (20°C). Seed of the American species benefits from stratification.

An interesting alternative method that works for most species is to remove the flower heads and float them upside down is shallow water over soil. They soon anchor themselves and produce new plants.

Cyperus alternifolius

Cyrtanthus Fire Lily, Scarborough Lily

These late summer- to autumn-flowering African bulbs form small to medium-sized clumps that may be divided in winter. The divisions can take a while to settle down to flowering so don't break up the clumps more regularly than every three years.

Fresh seed sown at 68–72°C (20–22°C) germinates freely but may take several years to flower.

Cyrtanthus o'brienii

Broom spp.

Genista aetnensis

Dahlia 'Apache'

Daphne odora 'Leucantha'

Davidia involucrata

Cytisus, Genista and Spartium Broom

Though pretty, brooms are weeds in many areas, so not surprisingly their seed germinates quickly if sown fresh at around 68°F (20°C). Stored seed will need soaking for 24 hours before sowing.

Hybrids and cultivars are best raised from heeled semi-ripe cuttings taken in late summer. Winter hardwood cuttings will also strike but can be very slow.

Types that are reluctant to strike may be grafted on to *Laburnum* seedling stocks in spring using a whip and tongue graft.

Dactylorhiza Marsh Orchid

Orchid seeds are minute and usually require specialist care so the simplest propagation method is division, which in the case of these terrestrial orchids is a straightforward procedure. It is best done in early spring with the divisions kept on the large side. They may be slow to recover but generally present no problems.

Dahlia

Dahlias are most commonly propagated by dividing the clumps of tubers when dormant. The tubers can be cut to provide more plants, though they rot rather easily. Spring cuttings of the vigorous shoots strike well, but cuttings often produce a few "blind" tubers that fail to shoot in the following season.

Dahlias can also be raised from seed and the dwarf bedding dahlias are usually seed-grown, being sown in spring, lightly covered, at 64–68°F (18–20°C).

Daphne

The species, such as *D. mezereum* and *D. cneorum*, are often grown from seed, but may be grown from semi-ripe summer cuttings or by layers. *Daphne genkwa* is raised from seed or root cuttings.

The common *D. odora* cultivars can be grown from cuttings but because of virus troubles many are produced by tissue culture, resulting in plants that are vastly superior in appearance and vigor. If you wish to grow cuttings, try to secure them from tissue-cultured bushes, which should be reasonably virus-free.

Davidia Dove Tree, Handkerchief Tree

Known for its impressively large, white flower bracts, this deciduous tree is raised from seed or late summer semi-ripe cuttings. The seed is best sown as soon as it is ripe and left outdoors to germinate in the spring. Otherwise, stratify it indoors and sow in spring. Either way keep the seed in moist soil because if dried it will take far longer to germinate. First-year seedlings or cuttings need frost protection.

Delphinium

Delphiniums may be grown from seed, divisions or cuttings. The seed must be fresh and is usually sown in early autumn or spring at 59–75°F (15–24°C). Germination is quite rapid. Avoid disturbing the seedlings any more than is absolutely necessary.

Divisions are made in late winter. Cuttings of the fleshy basal shoots are usually taken in early spring, and are among the few that do not strike well under mist. They do better in boxes kept in a bright but cool position. The similar-looking but annual larkspur (*Consolida ambigua*) is raised from seed.

Delphinium x *belladonna* 'Piccolo'

Desfontainea

Resembling some kind of exotic flowering holly, this Chilean and Peruvian shrub may be raised from seed or cuttings. Sow the seed, lightly covered, at 59–68°F (15–20°C) and once germinated keep the seedlings cool and moist. Softwood and semi-ripe cuttings taken during the warmer months will strike, though the percentages are often low.

Desfontainea spinosa

Deutzia

The species and the cultivars are propagated mainly by cuttings. Softwood cuttings taken in summer strike well, especially under mist. Alternatively, try hardwood cuttings outdoors over winter. The very bushy or low-growing forms sometimes self-layer.

Deutzia crenata 'Nikko'

Dianthus Pinks, Carnation

Dianthus will germinate freely from seed sown in spring at 70°F (21°C). Perennial forms may be grown from layers, or cuttings, which are commonly known as "slips." Taken at any time during the growing season these will strike reasonably quickly, but the roots are often poor and easily damaged. Take care not to damage the roots when transplanting.

Dianthus 'Far Cry'

Diascia Twinspur

Twinspurs include annuals and perennials. The annuals must be grown from seed, sown lightly covered at around 68°F (20°C), and the perennial species can be treated similarly. The perennial cultivars strike readily from softwood cuttings taken from spring to autumn.

Diascia barberae 'Ruby Fields'

Dicentra Bleeding Heart

The species may be raised from seed or cuttings. While stratification is often recommended for the seed it seems to germinate quite readily without it. Give it cool temperatures, from 55–64°F (13–18°C), and be prepared to wait up to a month. The selected forms are grown from divisions or cuttings in much the same manner as delphiniums. Cuttings of the basal shoots strike freely and are a quick way of increasing your stock.

Dicentra formosa

Dictamnus albus 'Roseus'

Dierama pendulum

Dietes iridioides 'Johnsonia'

Digitalis x *mertonensis*

Osteospermum barberiae

Diospyros lotus

Dictamnus Burning Bush

Known for its showy flowers and volatile oil, this woody-based Eurasian perennial may be propagated by seed or root cuttings. Sow the seed in spring at around 59°F (15°C) after a six-week stratification period or alternatively sow the seed when fresh in late summer and let it over-winter outdoors. Insert root cuttings in light soil in early spring.

Large clumps can be divided but are difficult to break up and may take several years to recover.

Dierama Wand Flower

Large clumps may be divided in late winter or sow the seed in late summer as soon as it is ripe. Cover the seed and keep at around 68°F (20°C). The seedling will take a couple of years to reach flowering age.

Dietes Fortnight Lily, Wild Iris

Because seedlings take a while to bloom, these perennials are usually propagated by simply dividing well-established clumps in late winter. The divisions quickly re-establish and will flower in the following season.

Digitalis Foxglove

The commonly grown species, such as *D. purpurea*, tend to be biennials or short-lived perennials and are often best raised from seed to keep the stock vigorous. Sow the seed uncovered, in early autumn or spring, at around 68°F (20°C).

Clumps that are reliably perennial can be divided in late winter or early spring, or the basal shoots can be used as cuttings.

Dimorphotheca and *Osteospermum* African Daisy, Veldt Daisy

These southern African daisies are quick-growing annuals, perennials and sub-shrubs. The annuals must be raised from seed, but the perennial types are also easily propagated by softwood tip cuttings taken any time they are available. Sow annual or species seed, lightly covered, at around 68°F (20°C).

Diospyros Persimmon, Sapote

American persimmon (*D. virginiana*) can be propagated by root cuttings taken in spring or from hardwood cuttings. Seeds of it and the date plum (*D. lotus*) will also germinate in spring at around 68°F (20°C) if previously exposed to cold, either wintered outdoors or stratified in a refrigerator. However, the best varieties of Japanese persimmon (*D. kaki*) are nearly always budded or grafted on to *D. virginiana* seedling rootstocks.

Dipsacus Teasel

Once used for teasing wool and other fibers, the feature for which these plants are most often grown — their interesting and intricate seedheads — is the obvious source of propagation material: seeds. Harvest the seeds from the dried heads in fall, before the birds do, and sow them in spring, lightly covered, at around 68°F (20°C). Seedlings should begin to appear within a couple of weeks.

Dipsacus fullonum

Dodonea Hop Bush, Ake Ake

These evergreen shrubs can be raised from seed—the purple-leafed forms come reasonably true to type—or from spring to autumn semi-ripe cuttings, which strike well and quickly, especially with bottom heat and mist. While there is less chance of rotting if the seed is removed from its papery capsule it will still germinate well if sown in spring, lightly covered, at around 68°F (20°C).

Dodonea boronifolia

Dombeya Wedding Flower

These rather tender, soft-wooded shrubs grow very quickly and flower from a young age. Semi-ripe spring to autumn cuttings strike well but trim the leaves to make the cuttings less unwieldy. Fresh seed germinates within two weeks at around 72°F (22°C) and is the best way to obtain large numbers of species plants. Double-flowered forms may be reluctant to set seed.

Dombeya cultivar

Dracaena Dragon Tree

Like some of the yuccas, dracaenas are well known for their ability to grow from "logs"—lengths of stem with buds. Treated like a large cutting, the log soon sprouts leaves and with time will develop a strong root system. Foliage heads can also be treated as cuttings and many dracaenas produce suckers and offsets that can be removed and grown on. The offsets sometimes have roots attached but more often need to be treated as cuttings.

Dracaena surculosa 'Florida Beauty'

Dracocephalum Dragon's Head

These hardy perennials are most easily propagated by dividing the rootstock as new growth commences in late winter or early spring. The divisions may take a little while to re-establish but are otherwise trouble-free. Keeping them moist and well-fed will ensure a quick recovery.

Dracocephalum ruyschianum

Dracunculus Dragon Arum

These impressive tuberous perennials are most easily propagated by dividing well-established clumps during their dormant period in winter or very early spring. The plants re-establish quickly and usually flower in the following summer.

Dracunculus vulgaris

Drimys winteri

Drimys Winter's Bark

These tough evergreens are most commonly propagated by cuttings. Softwood and semi-ripe cuttings taken from late spring until late autumn are usually successful but do far better under mist with bottom heat. If cuttings prove difficult, layers are a reliable if slow method.

Dryas octopetala

Dryas Mountain Avens

These small shrublets often self-layer, which is the easiest way to obtain a few new plants. Simply remove the rooted pieces and grow them on. If more plants are required, spring to autumn softwood and semi-ripe cuttings strike quickly and the seed germinates well if sown on light soil in spring at around 64°F (18°C).

Eccremocarpus scaber

Eccremocarpus Glory Flower

These South American climbers are easily raised from cuttings, layers or seed, which often self-sows and can be a nuisance. Sow the seed, lightly covered, at around 68°F (20°C) and it should be up within a couple of weeks. Softwood or semi-ripe cuttings will strike well at any time during the warmer months.

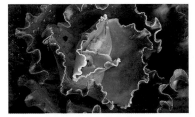

Echeveria 'Firelight'

Echeveria

By far the quickest method of propagation is simply to remove individual rosettes and grow them on. Sometimes they will have roots attached, other times they won't, but either way the roots will develop and the new plants soon settle down to steady growth. If you do not have a parent plant or want greater numbers, sow the seed, which germinates best at warm temperatures, around 77°F (25°C), and should not be covered.

Echinacea purpurea

Echinacea Cone Flower

Cone flowers can be propagated by seed, division or cuttings. The seed should be sown in early spring, lightly covered, at around 72°F (22°C). Divide established clumps just as they start to show signs of new growth in late winter or early spring, or use the strong spring basal shoots as cuttings.

Echinops battanicus

Echinops Globe Thistle

Although this genus includes annual, biennial and perennial species, most of those grown in gardens are perennial and may be propagated by seed, division or cuttings. Sow the seed uncovered, at around 68°F (20°C). Divide clumps between late autumn and early spring or use the strong basal shoots that develop in spring as cuttings. *Echinops* will also grow from root cuttings taken during winter.

Echium

This genus includes annuals, biennials, quick-growing perennials and a few shrubs. The annuals and biennials, such as viper's bugloss (*E. vulgare*), are easily grown from seed sown, lightly covered, at around 68°F (20°C). The perennials, like the shrubby species, are usually grown from cuttings because they do not produce a readily divisible clump. Rooted side-shoots can be removed, but most often it is necessary to use tip cuttings, which may be taken from spring to autumn. Often the strike rate is not high, but the plants produce so much cutting material that it seldom matters.

Echium candicans

Edgeworthia Paper Bush

These beautiful and unusual daphne relatives are most often propagated by semi-ripe cuttings taken in summer and autumn. Softer wood will strike but mist is essential as the young leaves quickly wilt. Timing is important with these plants and experience and a little luck to get good results.

Edgeworthia papyrifera

Elaeagnus Oleaster

These very hardy shrubs are usually propagated by cuttings: softwood and semi-ripe warm-season cuttings for the evergreens and winter hardwood for the deciduous. Sometimes they are grafted but this is seldom necessary. If species are required, sow the seed outdoors when ripe or stratify for 12 weeks then sow in spring, covered. It requires surprisingly warm temperatures, 68–77°F (20–25°C), to germinate well.

Elaeagnus pungens

Embothrium Chilean Fire Bush

Spectacular in bloom and hardy for protea family plants, these South American shrubs and small trees are not always easy to propagate. They can be raised from seed, which should be sown on a sphagnum-based mix, but are more often propagated by taking softwood and semi-ripe cuttings during the warmer months. The cuttings often strike well but fail to develop. Steadily cool, moist conditions suit them best.

Embothrium coccineum

Enkianthus

The best flowering and autumn foliage forms of these Japanese deciduous shrubs are propagated by cuttings. Members of the erica family, they grow readily from softwood and semi-ripe summer to autumn cuttings, especially under mist with a little bottom heat.

Enkianthus perulatus

Ephedra Mormon Tea, Joint Fir

These near-leafless shrubs, the source of several important drugs, are usually raised from the seed found in the berry-like fruits that follow the female flower cones. The seed usually germinate freely if sown, lightly covered, at around 68°F (20°C) in spring, though some of the hardier North American species may benefit from stratification.

Ephedra viridis

Epimedium x versicolor

Eranthis hyemalis

Erica 'Linton's Red'

Eriobotyra japonica

Eriophorum vaginatum

Eriostemon myoporoides 'Stardust'

Epimedium Bishop's Hat, Barrenwort

Although seed sown in autumn or spring will germinate readily, because many of the garden forms are cultivars and because they divide so readily, it is seldom necessary to sow seed. The clumps may be broken up at any time between late autumn and early spring but they re-establish most rapidly if divided just as the first shoots appear in late winter.

Eranthis Winter Aconite

Tuberous perennials, Eranthis may be divided or raised from seed. Sow the seed in autumn and leave to over-winter outdoors or stratify and sow, lightly covered, in spring. Seedlings will flower in two to three years.

Divide established clumps in autumn and keep the divisions on the large side.

Erica, Calluna and Daboecia Heath and Heather

These very adaptable plants may be propagated in several ways. The species grow rapidly from seed sown at 64°F (18°C) on a mix high in sphagnum moss. Most European types will naturally self-layer as they grow, and these layers may be removed and potted.

Cuttings are best taken from late summer through autumn. Take very tiny (1 in [2.5 cm] or less) cuttings from non-flowering side-shoots. A few of the southern African ericas are tricky, but cuttings of most will rapidly develop masses of fibrous roots, especially with mist and bottom heat.

Eriobotrya Loquat

Although easily raised from seed, the best fruiting forms are grafted on to seedling stocks, most often in early spring using a side graft. For grafting stock and species soak the seed for 24 hours before sowing at 64–75°F (18–24°C). Semi-ripe summer and autumn cuttings often strike quite well but the success rate can be rather unpredictable.

Eriophorum Cotton Grass

Like most grassy plants these sedges are easily divided and if broken up in spring as the new growth commences they will quickly re-establish themselves.

Eriostemon Waxflower

Cuttings are the most common method of propagating these attractive and long-flowering Australian shrubs. Softwood and semi-ripe cuttings may be taken whenever they are available and will strike readily, especially under mist with bottom heat.

Eryngium Sea Holly

Sea hollies may be annual, biennial or perennial. The annuals and biennials are, of course, propagated by seed. Sow it, lightly covered, in spring at around 64–68°F (18–20°C).

The perennials are usually propagated by root cuttings, which are taken in late winter or very early spring.

Eryngium causacicum

Erysimum Wallflower

Wallflowers may be annual or perennial. The annuals are quite frost hardy and even young seedlings will usually safely over-winter outdoors. The seed may be sown at any time from late summer until late spring, though it germinates best with relatively cool conditions—around 59°F (15°C). The perennials may be grown from layers, which the very dwarf types often develop naturally, or by softwood semi-ripe tip cuttings taken from spring to autumn.

Erysimum cultivars

Erythrina Coral Tree

These spectacular flowering trees have hard seeds that need scarifying and/ or soaking before sowing. The seed should be covered and germinates best at warm temperatures—around 77°F (25°C). Alternatively, take softwood or semi-ripe cuttings during the warmer months.

Erythrina caffra

Erythronium Dog's-tooth Violet

These lovely perennials may be divided when dormant but sometimes they react badly to root disturbance and should really only be broken up when the clumps become over-crowded.

However, as they usually set seed quite freely, propagation is not difficult. Sow the seed when ripe and leave to over-winter outdoors or stratify for eight weeks then sow in spring. It germinates at cool temperatures and should be only lightly covered.

Erythronium californicum

Escallonia

The species may be grown from seed sown at 70°F (21°C), but as most garden specimens are selected forms, cuttings are more commonly used. Softwood or semi-ripe cuttings taken at any time through the growing season usually strike quickly and develop without trouble.

Escallonia rubra var *macrantha*

Eschscholzia California Poppy

Often self-sowing only too readily, California poppies are now available in a wide range of colors and styles. The seed germinates best at around 68°F (20°C) on a fairly sandy mix and needs only a light covering of soil. The seedlings do not always transplant well and generally do best sown *in situ*.

Eschscholzia californica

155

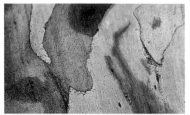

The bark of *Eucalyptus delegatensis*

Eucryphia moorei

Euonymus europeaus

Eupatorium cannabinum

Euphorbia pulcherrima

Euryops pectinatus and *E. tenuissimus*

Eucalyptus Gum Tree

Most species will grow readily from seed sown in spring or summer at 68–77°F (20–25°C). To avoid damaging the root system, transplant the seedlings at an early age. It is possible to strike cuttings of a few species, such as *E. camaldulensis*, but most are difficult and inclined to collapse even if they strike roots. Grafting of selected forms has been tried with limited success.

Eucryphia Roble de Chile, Pinkwood

Although the species may be raised from seed, those grown in gardens tend to be selected forms and usually originate from cuttings. Use softwood or semi-ripe tip cuttings taken from late spring to autumn. They can take quite a while to strike and the success rate is variable.

Euonymus Spindle Tree

Euonymus cultivars are usually propagated by cuttings and most methods work well. Try semi-ripe summer and autumn cuttings for the evergreen types and winter hardwood cuttings for the deciduous forms. Most are so easy, they will strike at any time and many of the deciduous forms will also grow from small tip cuttings taken in summer. Seed of the species germinates well but usually requires stratifying.

Eupatorium

Both herbaceous and shrubby species of these mainly North American natives are easily propagated. The herbaceous types may be divided when dormant, from late autumn to early spring and the shrubs will grow from softwood and semi-ripe cuttings. Also, seed may be sown in spring and often germinates more evenly if stratified for a few weeks before sowing.

Euphorbia

Euphorbia is among the largest plant genera, with some 2,000 species covering a huge range of plant types and sizes. Most are, however, very straightforward to propagate, growing easily from seed or stem cuttings. The requirements for the seed vary with the species, but sow it uncovered at around 68°F (20°C) and you should not be far wrong. Allow cuttings to dry, so the latex stops running, before inserting in the cutting mix. Apart from that, most types, even the large succulents, strike quite readily.

Euryops

These rather variable African shrubby daisies are most often propagated from semi-ripe cuttings taken in summer and early autumn, though they will strike from almost any unhardened wood in any season, especially with mist and bottom heat. The pretty silver-leafed *E. acraeus* is more often raised from seed, which germinate well, lightly covered, on gritty soil at around 68°F (20°C).

Exochorda Pearlbush

Often these Chinese deciduous shrubs produce suckers and if only a few extra plants are required these can be removed and grown on. Otherwise take summer to autumn semi-ripe cuttings or sow seed. The cuttings must be well-established by mid-autumn if they are to be overwintered outdoors.The seed germinates more evenly if stratified for a few weeks. Sow at around 68°F (20°C) with a light covering of soil.

Exochorda x *macrantha*

Fagopyrum Buckwheat

Now properly reclassified under *Polygonum*, according to most authorities, the common, fragrant-flowered *F. esculentum* (*P. fagopyrum*) differs from most species in that genus in being an annual. The seeds, which are often ground to make a flour-like meal, germinate best at temperatures over 68°F (20°C) and should be sown lightly covered. They usually sprout within 2–3 weeks.

Fagus sylvatica forma *laciniata*

Fagus Beech

Seedlings are seldom grown except to provide cheap hedging plants or as grafting stock. Stratify the seed for eight to 12 weeks and sow, lightly covered, at around 68°F (20°C).

Softwood to semi-ripe cuttings taken in early summer and kept under mist strike reasonably well, but most of the selected forms, such as the beautiful 'Zlatia,' are grafted on to seedling *F. sylvatica* stock.

Fagus sylvatica purpurea

Fallopia Lace Vine, Fleece Flower

These tough, twining or trailing perennials often grow only too well, self-sowing and covering large areas and anything that gets in their way. The plants may be raised from seed, softwood and semi-ripe spring and summer cuttings, or by removing and replanting small rooted pieces, preferably in early spring.

x *Fatshedera lizei*

x *Fatshedera*

This interesting and popular plant is a rare example of intergeneric hybridization and although it flowers it does not produce fertile seed. It is, however, easily propagated by taking semi-ripe cuttings at any time. Alternatively its rather sprawling habit and lengthy internodes make it a natural candidate for layering.

Fatsia Japanese Aralia

Semi-ripe *Fatsia* cuttings strike easily, but are large and unwieldy. Unless you want to grow a variegated cultivar, it is usually easier to sow the seed, which is produced in abundance and germinates very freely. It does not need to be covered and germinates best at around 68°F (20°C).

Fatsia japonica 'Variegata'

157

Felicia angustifolia

Ficus carica

Filipendula vulgaris

Forsythia intermedia

Fothergilla gardenii

Francoa ramosa

Felicia Kingfisher Daisy

Annual kingfisher daisies often self-sow, so it is no surprise that the seed germinates quickly if sown, lightly covered, at around 64°F (18°C). The perennial species, hybrids and cultivars, such as the very popular 'Santa Anita' and 'San Gabriel,' strike quickly from softwood tip cuttings taken at any time. The very soft bark is easily stripped, so take care when preparing the cuttings.

Ficus Fig

This large family of shrubs, climbers and trees is notable for its latex sap and a tendency to develop aerial roots. Although they are not difficult to raise from seed (sown uncovered at around 72–77°F [22–25°C]), most cultivated plants are cutting-grown or raised from aerial layers. Those with small leaves develop quickly from spring to softwood or semi-ripe cuttings, but because the large-leafed types make unwieldy cuttings, they are often aerially layered. Indeed, the rubber tree (*F. elastica*) is the classic subject for demonstrating aerial layering.

Filipendula Meadowsweet

Easily lifted and divided when dormant from late winter to early spring, these tough perennials may also be raised from seed, though many of the best forms are cultivars that will not reproduce true to type. The seed germinates well at moderate temperatures when only lightly covered.

Forsythia

The species may be grown from seed but, as most are selected forms, cuttings are more common. Small softwood or semi-ripe cuttings will root rapidly during the growing season, or hardwood cuttings may be struck over winter. As forsythias are plants with pithy stems, take hardwood cuttings near a node to prevent the pith drying or rotting.

Fothergilla

These deciduous shrubs have plume-like flower spikes and are usually propagated by taking summer softwood cuttings. These strike best under mist with mild bottom heat. If you do not have a misting system, take slightly firmer cuttings that will be less inclined to wilt. Failing that, winter hardwood cuttings will strike but they are slower.

Francoa Bridal Wreath

Seed, division and cuttings are all suitable methods for these perennials. The seed is best sown in spring and will germinate at moderate temperatures. Established clumps can be divided in late winter or early spring or the individual rosette-like foliage whorls can be removed from the clump and treated as cuttings.

Fraxinus Ash

The species may be raised from seed but will not flower until several years old. Seed cultivation is important, though, because cultivars are usually grafted onto seedlings of the same species.Sow the seed in spring after stratifying it for around 12 weeks. It germinates best lightly covered at 64–72°F (18–22°C).

Softwood or semi-ripe cuttings taken in summer have a strike rate of around 50 percent. Winter hardwood cuttings also strike moderately well, but most of the cultivars seen in gardens have been grafted.

Fraxinus excelsior

Freesia

Freesias are corms that over the years have been extensively hybridized so that modern seedling strains are quite reliably true to type. Sow the seed at around 68°F (20°C) after soaking it overnight. Keep covered.

Named cultivars must be increased vegetatively and this is just a matter of waiting for them to multiply naturally. There is not much you can do to speed up the process in cool areas, so if freesias do not do well outdoors in your climate, keep to seedlings or buy your corms.

Freesia cultivar

Fremontodendron Flannel Bush

These attractive evergreen shrubs from California, western Arizona and northern Mexico have felted maple-like foliage and showy golden-yellow flowers. The hard black seeds are the best propagation option. They need scarifying and/or soaking to soften them and they should be sown, covered, at 64–75°F (18–24°C). The seedlings are inclined to collapse, usually through damping off, and need spraying with a fungicide.

Fremontodendron californicum

Fritillaria Fritillary

Beautiful and often very rare in the wild, fritillaries have a reputation for being hard to grow. They tend to be slow to multiply naturally and difficult to hurry along. Consequently, seed is the most reliable propagation method. Stratify for at least eight weeks and sow lightly covered at around 64°F (18°C).

If you are very lucky, some of your fritillary clumps may become large enough to break up. This can be done in autumn and winter, but you may be better off leaving well alone—they tend to resent disturbances.

Fritillaria oliveri

Fuchsia

The species or your own hybrid crosses can be raised from seed. While the seeds can be difficult to separate from the fruit pulp, it germinates freely at 70°F (21°C). Avoid completely covering fuchsia seed, as it needs some light.

Fuchsia hybrids and cultivars are grown from softwood or semi-ripe cuttings taken during the growing season. These strike quickly, especially with mist and bottom heat.

Fuchsia 'Violet Lee'

Gaillardia cultivar

Gaillardia Blanket Flower

This genus is another with annual, biennial and perennial species. As always, the annuals and biennials are raised from seed, sown uncovered at around 68°F (20°C), while you can propagate the perennials by division or cuttings of vigorous young shoots. Divide the clumps in late winter or early spring. Root cuttings taken in early winter will come into growth during spring.

Galanthus nivalis

Galanthus Snowdrop

Snowdrop bulbs are small and largely unsuited to special treatments such as scooping. Fortunately they often increase fast enough naturally to build up stocks quite quickly. Seed is another method and is usually produced in good quantities. Stratify for six weeks and sow, covered, with cool conditions—around 59°F (15°C).

Galega officinalis

Galax Beetleweed, Wandflower

The sole species in this genus is a beautiful foliage groundcover for shady areas that also offers delicate sprays of white flowers in summer and bronze winter foliage. It is an evergreen rhizomatous perennial that is easily divided in late winter and early spring as it starts into new growth.

Galega Goat's Rue

Both commonly grown species may be propagated by seed or division. Seed sown in spring at around 68°F (20°C) and lightly covered should be up within a couple of weeks. Established plants may be divided at any time during the late autumn to early spring period but later divisions tend to re-establish more quickly.

Galtonia candicans

Galtonia Summer Hyacinth

If many plants are required sow the seed, which is freely produced. It germinates quickly if sown in spring at around 68°F (20°C). For just a few plants or to propagate selected forms, simply remove some of the small offset bulbs from around the edge of the clump. It is rarely necessary to break up established clumps.

Gardenia augusta

Gardenia

Softwood or semi-ripe cuttings taken in summer have a high strike rate, especially with mist and bottom heat. The species can also be raised from seed sown at around 72°F (22°C). Many gardenias, especially *G. thunbergia*, produce large seedpods and the seed is easily harvested.

Garrya Tassel Tree, Silk-tassel Bush

These beautiful winter-flowering shrubs and small trees are not always easy to propagate without sophisticated equipment. Semi-ripe tip or sideshoot cuttings, best taken with a heel, will strike, but mist and bottom heat are usually needed to ensure there is good and reasonably quick root formation.

Garrya elliptica 'James Roof'

Gaultheria Wintergreen, Snowberry

These shrubs vary somewhat in their growth habit. The stoloniferous types can be propagated by division in late winter; the more bushy forms by cuttings or layers; all may be raised from seed. The seed is best sown on a sphagnum-based mix, uncovered, at around 64°F (18°C). Softwood and semi-ripe cuttings taken during the warmer months strike freely, especially with mist and mild bottom heat.

Gaultheria crassa

Gaura

This North American genus includes annuals, biennials and perennials, though only the perennials are generally cultivated. They are most often propagated by dividing established clumps in late winter or early spring or by taking small cuttings of non-flowering shoots from spring to autumn. Late cuttings need to be either well established or kept protected if they are to survive harsh winters.

Gaura lindheimeri

Gazania Treasure Flower

Although perennial gazanias can be propagated by softwood cuttings or by carefully breaking up established clumps, most plants are raised from fresh seed. Sow the seed in spring, lightly covered, at around 68°F (20°C). Use cuttings to perpetuate any particularly attractive forms.

Gazania hybrids

Gelsemium Carolina Jessamine

This is a vigorous climber that often produces masses of fragrant yellow flowers. It can be raised from seed, but most plants are propagated by semi-ripe cuttings. These can be taken at any time during the growing season and while sometimes slow to strike, they are reliable.

Gelsemium sempervirens

Gentiana Gentian

Gentians can be very tricky to cultivate, but most will germinate quite readily from seed. Stratify the seed for a few weeks and sow it, lightly covered, at around 64°F (18°C).

Gentians that spread and shoot from the rootstock can be divided in winter if you are very careful and keep the divisions large. Others will grow from softwood cuttings of the non-flowering stems.

Gentiana septemfida

Geranium maderense

Gerbera hybrid

Geum chiloense

Ginkgo biloba

Gladiolus hybrid

Gleditsia japonica

Geranium

The true geraniums, not to be confused with pelargoniums, often multiply only too well, sometimes becoming invasive weeds. They produce large quantities of seed, but unless you want to grow the species, division and cuttings of the young shoots are preferable techniques. Established clumps can be broken up at any time and even small divisions develop quickly. Take cuttings throughout the growing season. If you want to sow the seed, cover it lightly and give a temperature of around 68°F (20°C).

Gerbera Transvaal Daisy

Seed sown in spring or summer germinates well at around 72°F (22°C). The seedlings should be transplanted when quite small to avoid too much root disturbance. Remove basal shoots in spring. If these can be removed with a few roots attached, so much the better, but if not simply treat them as cuttings until the roots develop. Large clumps can be divided with care.

Geum Avens

This genus includes clump-forming perennials that can be divided in late winter to early spring and sub-shrubs that are easily propagated by softwood or semi-ripe cuttings in late spring and summer. Many of the perennials also strike from softwood cuttings of the young non-flowering shoots. The species may be raised from seed sown uncovered at around 68°F (20°C). Stratification often yields higher germination rates.

Ginkgo Maidenhair Tree

Stratified seed sown, covered, at around 72°F (22°C) germinates well, but you will not know the sex of the seedlings until they mature. Female trees produce fleshy fruits that have an unpleasant smell, especially when trodden on. Softwood or semi-ripe summer cuttings strike well and if you know the sex of the tree from which the cuttings were taken you will know the sex of your new trees. Males appear to strike roots more freely than females.

Gladiolus

Gladiolus corms usually multiply very well by themselves, producing many small cormels around each of the large corms. These can be grown on and will flower after two or three years depending on the type. Fancy hybrids often have large corms that can be cut to produce extra plants, but the risk of rot is high and the speed of natural increase makes it largely unnecessary.

Gleditsia Honeylocust

While winter hardwood cuttings strike quite well, most garden cultivars, such as 'Sunburst,' are grafted, usually on to *G. triacanthos* seedling stock. The bean-like seeds germinate well at around 68°F (20°C) but scarify and soak before sowing.

Globularia Globe Daisy

These small perennials or shrublets are often raised from seed, which germinates well at cool temperatures, around 59°F (15°C). Some species can be divided in late winter; others will form natural layers; and others have stolons that can be removed and grown on.

Globularia sp.

Gloriosa Glory Lily, Flame Lily

These climbing lilies are tuberous perennials that can be divided at the end of the growing season. They are often grown in containers and repotting time is usually the best time to divide the tubers, which will redevelop from even quite small pieces provided each has an eye. Glory lilies frequently produce seed that can be collected when ripe in the autumn and sown in spring. It germinates at 68–74°F (20–23°C) and the seedlings flower in around three years.

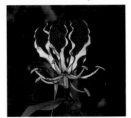

Gloriosa rothschildiana

Gordonia

Related to the camellias, these evergreen shrubs and small trees may be propagated from seeds or cuttings. If lightly covered and kept at around 68°F (20°C) the seeds germinate in two to four weeks. Softwood and semi-ripe cuttings taken during the warmer months have a rather variable strike rate but are usually quite reliable if mist and mild bottom heat are available.

Gordonia axillaris

Grevillea Spider Flower

Grevilleas grow well from summer and autumn softwood and semi-ripe cuttings, though the strike rate varies with the species and cultivar. Mist and bottom heat improve the strike rate and speed of root formation. The species may also be raised from seed. Soak the seed for a few hours and sow it, covered, at around 72°F (22°C). The seedlings of some species are inclined to damp off or suddenly collapse. Others, such as silky oak (*G. robusta*), germinate well and grow rapidly.

Grevillea 'Robyn Gordon'

Griselinia

Although *Griselinia* seed germinates well, most garden plants are raised from semi-ripe cuttings. These can be taken at any time and will usually strike within a month, especially with mist and bottom heat. The roots of the freshly struck cuttings can be brittle, so take care when transplanting them.

Griselinia littoralis

Gunnera

Spectacular plants and rapid growers, these strong rhizomatous perennials are easily propagated by seed or division when dormant in winter. The sheer physical effort required to lift and divide the heavy rhizomes of the South American species means seed is far easier to handle; sow it in damp soil in spring at 59–68°F (15–20°C) and the seedlings should start to appear within a couple of weeks. The tiny New Zealand species are easily divided.

Gunnera prorepens

Gymnocladus chinensis

Gypsophila paniculata 'Bristol Fairy'

Haemanthus coccineus

Hakea laurina

Halesia tetraptera

Halimium lasianthum

Gymnocladus Kentucky Coffee Tree

Although the seeds found in the large pods of *G. dioica* were once used as a coffee substitute this large tree, the only commonly grown species, is now treated strictly as an ornamental. The seeds, which appear on female trees only, are worth having, however, as they are the main method of propagation. Sow them outdoors in autumn or stratify them for 8–12 weeks before sowing in spring at around 68°F (20°C). Cover the seed and it should start to germinate within two weeks.

Gypsophila Baby's Breath

The annuals, species and certain hybrids may be grown from seed, which should be as fresh as possible. The seed germinates in about 12–20 days if sown, uncovered, at around 72°F (22°C). Selected forms are usually grown from cuttings, which strike quickest in late summer but may be taken at any time strong non-flowering shoots are available. Established clumps can sometimes be divided but often the divisions lack vigor.

Haemanthus Cape Tulip, Blood Lily

These spectacular bulbs from southern Africa can be divided but as they may take a considerable time to form a large clump they are often raised from seed. Sow in spring, lightly cover the seed and keep it warm—around 72–77°F (22–25°C) is best. The seed can take a while to germinate and the percentage is often low. The bulbs will respond to treatments such as scooping and scoring but the bulblet yield is usually low.

Hakea

Hakea seeds often have a very hard coat and benefit from scarifying and/or soaking. They germinate in two to four weeks at 64–77°F (18–25°C) and should be covered. Semi-ripe cuttings taken in summer to early autumn strike quite well with mild bottom heat but some species are inclined to simply form calluses without producing roots.

Halesia Silverbell, Snowdrop Tree

These deciduous shrubs and trees are most often raised from seed. Although it is capable of germinating if sown in a greenhouse when ripe, stronger plants may be produced and more even germination is obtained if seed is stratified for about eight weeks then sown in spring at around 68°F (20°C).

Halimium Sun Rose

Closely related to the rock roses (*Cistus*), these small evergreen shrubs are generally raised from softwood and semi-ripe cuttings taken during the warmer months. The cuttings strike quickly and well but layering, which sometimes occurs naturally, may be less trouble if only a few plants are required.

Hamamelis Witch Hazel

Witch hazels are cool-climate plants, so it is not surprising that the seed should be stratified before sowing. Give it about 12 weeks in the refrigerator then sow it, lightly covered, at around 64°F (18°C). Selected forms may be grafted but will usually grow well from softwood or semi-ripe cuttings taken from early summer to early autumn.

Hamamelis x *intermedia* 'Jelena'

Hardenbergia Coral Pea

This Australian genus of scramblers and climbers presents no propagation difficulties. The seed germinates freely at around 68°F (20°C) if soaked for a few hours before sowing. Semi-ripe cuttings can be taken at any time and they strike quickly. Although I haven't tried this method, I presume that layers would also form roots.

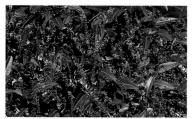
Hardenbergia violacea

Hebe

While most hebes grow easily from spring-sown seed, cuttings are the most common propagation method, even for the true species. Soft and semi-ripe cuttings taken during the growing season strike quickly, though whipcord hebes may prove difficult. Try sowing the seed or dropping, stooling or layering if you have difficulties with cuttings of these types.

Hebe x *andersonii* 'Variegata'

Hedera Ivy

Ivies are vigorous plants that are remarkably easy to propagate. Although they often self-sow, the easiest method of propagation is simply to dig up a few naturally formed layers. Failing that, layer a few branches yourself and wait for the roots to form. Or if you want still more plants, take semi-ripe cuttings in summer or early autumn.

Hedera canariensis 'Variegata'

Helenium Sneezeweed

This American genus includes annuals and biennials, which are raised from seed, and perennials, which can also be propagated by division and in some cases by taking cuttings of the basal shoots. Sow the seed in spring to early summer, uncovered, at around 68°F (20°C). Divide the perennials when dormant, preferably in late winter so they can re-establish quickly, and take cuttings of non-flowering shoots in spring and early summer.

Helenium autumnale 'Butterpat'

Helianthemum Sun Rose, Rock Rose

Sun rose seed germinates in two to three weeks if sown lightly covered at around 72°F (22°C). Softwood or semi-ripe cuttings can be taken throughout the growing season, striking quickly and reliably. Occasionally, stems near the ground will self-layer.

Helianthemum nummularium

Helictotrichon sempervirens

Heliopsis helianthoides

Heliotropium arborescens

Helleborus orientalis

Hemerocallis cultivar

Hepatica nobilis

Helictotrichon Oatgrass

The grasses of this genus, best known for it blue-leafed species and cultivars, are best propagated by division to retain the striking color of the best forms. Established clumps can be broken up at any time but late winter and early spring divisions will have the longest growing season available to re-establish themselves.

Heliopsis Orange Sunflower, Ox Eye

These tough perennials are most often propagated by dividing established clumps when they are dormant. Alternatively, seed can be sown in spring and will germinate quickly at around 68°F (20°C), though cultivars will not come true to type from seed. Many forms can also be grown by taking spring to early summer cuttings of vigorous, non-flowering basal shoots.

Heliotropium Heliotrope, Cherry Pie

Heliotropes are propagated from seed or by taking cuttings. Seed sown in spring, indoors, at around 68°F (20°C) will germinate within a couple of weeks. Cuttings of non-flowering side-shoots strike quickly. Autumn cuttings over-wintered under cover and regularly pinched back will make good bushy plants for the following summer.

Helleborus Hellebore, Lenten Rose

Most hellebores, apart from the cultivars, are raised from seed, which can be frustrating because although seedlings often pop up in the garden, they can be difficult to germinate under controlled conditions. Most species benefit from stratification but may need two periods. Consequently, it is often easier to sow the seed in the autumn and leave it to fend for itself. Sometimes the seedlings will appear in the spring, sometimes the following spring, sometimes never. Established clumps can be divided after flowering or when dormant, but this is a slow method of increase and not suitable for all types.

Hemerocallis Daylily

Daylilies are propagated by seed or division. If you are producing your own hybrids or are not too worried about getting exactly the right colors, sowing stratified seed at around 68°F (20°C) can quickly yield a huge number of plants. Hybrids and cultivars are perpetuated by dividing established clumps during winter.

Hepatica Liverleaf

These small perennials, by far the best known of which is *H. nobilis*, are usually propagated by division. Divide at any time during the cooler season dormant period up until the first leaves start to push through.

Hesperis Damask Violet

These sweetly scented biennials or short-lived perennials are usually raised from seed, which is best sown directly where it is intended to grow outdoors. The seed can be sown, uncovered, in autumn or spring in mild winter areas, but where the winter frosts are severe it is best left until spring.

Heterocentron elegans

Heterocentron Spanish Shawl

This genus of low, spreading perennials or shrubs is mainly propagated by division or cuttings. Most species form natural layers that can be removed with roots attached for growing on and those with a more perennial habit can be divided into large clumps in the cooler months. Tip or stem cuttings will strike quickly at any time.

Heuchera micrantha var *pacific*

Heuchera Coral Bells

Although *Heuchera* seed germinates freely if sown uncovered at around 64°F (18°C), most garden plants are hybrids or cultivars that must be propagated vegetatively. Division is the technique to use and established clumps can be broken up at any time from late autumn until early spring.

Hibbertia Guinea Gold Vine

These evergreen shrubs usually have lax stems and range from small ground covers through shrub to quite large climbers. They are most often propagated from softwood and semi-ripe cuttings taken during the warmer months. Many species will also layer easily, sometimes naturally.

Hibbertia aspera

Hibiscus

Most shrubby hibiscus grow well from seed sown at 73°F (23°C), but when it comes to vegetative propagation, distinctions must be made between the various types.

The tropicals are easily grown from semi-ripe cuttings during the growing season; young plants should be over-wintered indoors.

The hardy deciduous types are propagated from semi-ripe cuttings during the growing season, hardwood cuttings over winter, or by breaking up old clumps. Annuals must be grown from seed: scarify and/or soak before sowing at around 72°F (22°C).

Hibiscus trionum

Hippeastrum Amaryllis

Although *Hippeastrum* bulbs are large, they do not always respond well to severe treatments, such as scooping. The usual method of increase is natural offsets, which, of course, reproduce true to type, but if you simply want more bulbs, sow seed. Sow it lightly covered at around 72°F (22°C). Germination may take a while and the seedlings will not flower until they are at least two years old, but the method is reliable.

Hippeastrum 'Tangerine'

Hippophae rhamnoides

Hoheria populnea 'Alba Variegata'

Hosta decorata var *normalis*

Hovenia dulcis

Hoya carnosa 'Picta'

Hyacinthoides sp.

Hippophae Buckthorn

Buckthorns are vigorous deciduous shrubs and small trees that can be invasive in some areas. Not surprisingly then they are easily propagated, either by spring-sown seed or semi-ripe summer cuttings. The seeds should be stratified for 6–8 weeks before sowing at around 68°F (20°C). They should be covered and will start to germinate within three weeks.

Hoheria Lacebark

These mainly autumn-flowering New Zealand trees are members of the mallow family and in addition to flower clusters they often have attractive foliage and peeling bark. The species may be raised from lightly covered spring- or summer-sown seed kept at around 68°F (20°C), but the cultivars must be propagated vegetatively by taking semi-ripe cuttings of non-flowering shoots in late summer and autumn.

Hosta Plantain Lily

Most garden hostas are hybrids or cultivars that must be propagated vegetatively, which is very easily done by dividing established clumps in late winter or very early spring. It is sometimes claimed that the divisions "sulk," but I have always found them quick to recover. If you are keen on the species, the seed germinates well if sown lightly covered at around 68°F (20°C). Stratification often helps improve the germination rate.

Hovenia Raisin Tree

The far more commonly grown of the two species in this genus is an East Asian deciduous tree, the pedicels of which are edible and partially enclose the fruit. Because the seed needs scarification with a mild acid and the fruiting qualities of seedling trees are variable, it is usually propagated by softwood and semi-ripe summer and autumn cuttings. These may be slow with a variable strike rate but mist and bottom heat help considerably.

Hoya Wax Flower

The seeds of most *Hoya* species germinate quickly if sown at warm temperatures. The hybrids and cultivars should be propagated by softwood or semi-ripe cuttings taken during the warmer months. The waxy leaves and semi-succulent nature of hoyas gives them a moisture reserve, so the cuttings should be kept just moist—too wet and they will rot.

Hyacinthoides Bluebell

Bluebells usually multiply quickly enough if left to themselves. Just leaving a patch for two or three years then lifting it should yield plenty of young bulbs. If you need large numbers, seedlings will bloom within two years of sowing and the seed germinates well at around 64°F (18°C).

Hyacinthus Hyacinth

Hyacinths are the prime candidates for bulb scooping or scoring. These methods demand well-controlled conditions but are a quick way of dramatically increasing your stock of bulbs. Natural multiplication is slower, though very reliable. Hyacinth seed germinates well at around 68°F (20°C) but the seedlings take at least three years to produce their first flowers.

Hyacinthus orientalis cultivar

Hydrangea

Hydrangeas grow well from cuttings. Tip cuttings taken from non-flowering stems in spring and summer strike quickly and are usually well established by winter. Winter hardwood cuttings also strike well but may be slower to become established. Because the stems are pithy, cut just below a node to minimize the length of hollow stem that may develop.

Hydrangea macrophylla var *normalis* 'Blaumiese'

Hymenosporum Australian Frangipani

This fragrant-flowered Australian evergreen shrub or tree may be raised from seed or cuttings. The seed will germinate in two to three weeks if lightly covered and kept at around 72°F (22°C). The seedlings grow quickly and need pinching back to keep them compact. Softwood and semi-ripe cuttings taken during the warmer months strike well, especially with mist and bottom heat.

Hymenosporum flavum

Hypericum St. John's Wort

Hypericums, especially those with underground runners, often grow too well and keep coming up where they are least wanted. New plants can be started from runners, divisions or seed sown, covered, at around 68°F (20°C), but most hypericums are raised from softwood or semi-ripe cuttings. These can be taken at any time during the growing season and usually strike quickly.

Hypericum perforatum

Hypoestes Freckle Face, Polka-dot Plant

Usually treated as annuals and raised from spring-sown seeds, these plants are really perennials and evergreen shrubs that may also be propagated by taking softwood and semi-ripe cuttings during the warmer months. If you are sowing the seed, cover lightly and keep the temperature over 68°F (20°C).

Hyssopus Hyssop

Perennials or small shrubs, these popular herb garden plants may be raised from seed, cuttings and in a few cases by division. Lightly covered spring-sown seed kept at 59–70°F (15–21°C) will germinate in two to three weeks; softwood cuttings of non-flowering shoots taken from spring to autumn strike quickly; and the low, spreading forms can sometimes, with care, be divided over winter.

Hypoestes phyllostachya

169

Iberis sempervirens

Idesia polycarpa

Ilex aquifolium 'J.C. van Tol'

Impatiens walleriana

Incarvillea delavayi

Indigofera cytisoides

Iberis Candytuft

Annual and biennial candytuft is raised from seed that may be sown *in situ* but which is often best started under cover. Lightly covered and kept at around 68°F (20°C) it will germinate in two to three weeks. Perennial candytuft may be raised from seed too, which does not need covering and germinates at slightly lower temperatures. The perennials may also be divided in late winter or take cuttings from non-flowering stems during the growing season.

Idesia Wonder Tree

Known for its magnificent show of bright red fruit in winter while devoid of leaves, this deciduous tree is often raised from seed: stratify for a few weeks before sowing in spring at around 68°F (20°C). Particularly good forms may be perpetuated from semi-ripe cuttings taken in summer and early autumn. These strike well under mist with mild bottom heat.

Ilex Holly

Most garden hollies are hybrids or cultivars and are propagated by softwood or semi-ripe cuttings taken from late spring to mid-autumn. The species may be raised from seed, which should be stratified and sown, covered, at around 68°F (20°C). The germination time is very variable and can be as much as six months. Also, as hollies are often dioecious, it is not possible to know the sex of the seedlings until they mature.

Impatiens Busy Lizzie, Balsam

Bedding impatiens are usually raised from seed, which should be sown lightly covered at 72–77°F (22–25°C). The shrubbier types, such as the New Guinea hybrids, are also often seedlings, but they propagate very easily from cuttings, too. Any tip growth can be used as a cutting and will strike within a few days if taken during the warmer months.

Incarvillea

These late spring and early summer flowering perennials form clumps that are easily divided in late winter or early spring. If greater numbers are required, sow seed in early autumn or spring. It may be slow to germinate but only requires cool to moderate temperatures and a light soil covering.

Indigofera Indigo

The seeds of these perennial, shrubs and small trees have a hard coating and often need both scarifying and soaking. Those of most species will then germinate after a couple of weeks at 68°F (20°C). Clump-forming types may be broken into generous divisions in early spring and the shrubs will usually grow from root cuttings or winter hardwood cuttings, though they can be slow to strike.

Inula

Vigorous perennial daisies, often with strong rhizomes, these plants may be divided when dormant and will also grow from root cuttings taken in autumn. The seed germinates well in autumn or spring, lightly covered at around 68°F (20°C); autumn-sown seedlings may need winter protection.

Inula hookeri

Iochroma

These soft-wooded shrubs or small trees of the potato family grow readily from softwood and semi-ripe cuttings taken during the warmer months. Their large, soft leaves make them prone to wilting, so mist is a definite advantage. Pinch the tips from the young plants to encourage bushiness.

Iochroma grandiflora

Ipheion Spring Starflower

Clumps of these small bulbs are easily divided at almost any time, especially autumn and winter, and will quickly re-establish themselves. They also produce countless bulblets that can be grown on as new plants.

Ipheion uniflorum

Ipomoea Morning Glory

This is a very diverse genus, occurring in a wide range of forms, perennials, shrubs, trees and most notably climbers, some of which are annuals. The annuals are, of course, raised from seed: scarify and/or soak and sow at reasonably warm temperatures—over 68°F (20°C). Divide those with tuberous roots after flowering or when at their most dormant and raise the woodier types from semi-ripe cuttings taken during the growing season. Some species produce runners that can be lifted and grown on and other than the annuals most can be layered.

Ipomoea cairica

Iris

Iris roots may be fibrous, rhizomatous or bulbous and their propagation varies accordingly. Fibrous-rooted types, such as *I. unguicularis* and the Japanese irises, can be broken up at any time, usually when they are most dormant. The rhizomatous bearded irises are best lifted and divided in summer, preferably soon after flowering. Bulbous irises multiply fairly quickly by themselves, and if lifted when dormant will usually provide a good crop of young bulbs. The species can be raised from seed, which usually germinates more evenly if stratified for at least six weeks. Lightly cover the seed and keep it at around 68°F (20°C).

Iris ensata 'Oriental Glamour'

Itea Sweetspire

Found mainly in North America, this genus of deciduous and evergreen shrubs is known for its graceful panicles or racemes of small creamy white flowers, which in some species are fragrant. Semi-ripe cuttings taken in summer are the usual propagation method. They are sometimes reluctant to strike, though mist and bottom heat will speed them along.

Itea ilicifolia

Ixia paniculata

Jacaranda mimosifolia

Jasminum mesnyi

Juglans ailanthifolia

Juncus effusus

Ixia

Clumps of these bulbs may, of course, be broken up once well established. However, they are often rather slow to develop into a large clump and consequently seed can be a better option. If lightly covered and kept at around 72°F (22°C) the first seedlings should start to appear within two weeks.

Jacaranda

Jacarandas are spectacular trees and their seeds germinate very easily, especially if they are soaked for a few hours before sowing and then kept warm, at around 77°F (25°C). Semi-ripe cuttings taken in summer and autumn strike well.

Jasminum Jasmine

Jasmines can be grown from seed, but it is far more common to take semi-ripe cuttings in summer and autumn. While the strike rate varies with the type, most are quite easy. Established plants often send out long runners at ground level. These may strike roots, in which case the runners may be lifted and cut up for growing on. If you wish to try sowing seeds, keep them at around 72°F (22°C) and cover lightly.

Jeffersonia Twinleaf

These graceful woodland perennials may look rather soft and tender but they are in fact very hardy, though they resent root disturbance. This makes them difficult to propagate by division, but it can be done just as the plants are starting into growth in late winter and early spring. Use large divisions and be prepared to wait a while for them to re-establish. Seed is a safer method. Stratified for six weeks and sown, covered, at 59–68°F (15–20°C) in spring, the seedlings should appear within three weeks.

Juglans Walnut

Walnuts often self-sow and if you are considering them as ornamental trees only then seed is definitely the easiest propagation method. Stratification is usually necessary and may take up to 12 weeks. Because the seedlings resent root disturbance the seeds should be sown individually in pots or where they are intended to grow. The best fruiting forms will not come true to type from seed and are usually grafted on to seedling stocks. Some species can also be raised from winter hardwood cuttings.

Juncus Rush

Rushes are very easily propagated by division and may be broken up with complete success at any time, especially in spring. While the seed germinates freely, it is seldom used as most garden plants are cultivars.

Juniperus Juniper

Species junipers may be raised from seed, but garden plants are mostly propagated by cuttings. Use well-ripened wood at any time of the year or take softwood cuttings under mist in summer. Juniper cuttings definitely benefit from being taken with a heel.

Juniperus x *media* 'Blue Cloud'

Justicia

Popular as houseplants or outdoors in mild gardens, these perennials and shrubs strike quickly and reliably from softwood and semi-ripe cuttings taken during the warmer months. The species may also be raised from seed: sow, uncovered, at around 68°F (20°C).

Justicia chrysostephana

Kadsura

The only commonly grown species in this genus is an evergreen climber native to Japan and Korea. It is cultivated for its fragrant flowers and, if male and female plants are grown, for its bright red berries, which yield seeds that are seldom sown because gardeners prefer to know the sex of their plants. Consequently, warm season softwood and semi-ripe cuttings are preferred and these strike quite readily, especially under mist.

Kalanchoe

Kalanchoes are well known for forming small plantlets along their leaf margins. These eventually fall and strike root where they land. You can speed up the process by gently rubbing them off the parent leaf and "sowing" them on potting mix, or just pin down the whole leaf. Semi-ripe cuttings of the more succulent species are easily struck from spring to autumn. The popular flaming Katy (*K. blossfeldiana*) is usually treated as an annual and raised from seed sown, lightly covered, at around 72°F (22°C).

Kalanchoe x *blossfeldiana*

Kalmia Mountain Laurel, Sheep Laurel

All species will grow from seed sown in spring at around 59–68°F (15–20°C) on a mix high in sphagnum moss. If the seed is not fresh it may benefit from stratifying for about 12 weeks. Kalmias will grow from softwood cuttings, but the results are unpredictable and the usual method for the popular *K. latifolia* is layering, which is a time-consuming process. Sometimes, established plants will produce basal suckers, but not often. In recent years, tissue-cultured plants have become common.

Kalmia latifolia

Kerria

Softwood and semi-ripe cuttings taken from late spring to autumn strike very quickly, especially with mist and bottom heat. The double-flowered cultivar, 'Pleniflora,' is far more widely grown than the single-flowered *K. japonica*, but if you wish, the species may be raised from cuttings or seed.

Kerria japonica 'Pleniflora'

Knautia macedonia cultivar

Knautia

Closely related to *Scabiosa*, this genus of annuals and perennials is represented in cultivation by just a few of the perennial species. They can be raised from spring-sown seed, which sometimes benefits from stratification, but are most often propagated by taking softwood basal cuttings of non-flowering stems in spring and summer.

Knightia excelsa

Knightia New Zealand Honeysuckle, Rewa Rewa

Seed is the usual method for propagating this New Zealand member of the protea family. Gently scarify or nick the seed and sow at around 68°F (20°C) with a light covering of soil. Seedlings may take a while to appear and should be treated with a fungicide to prevent damping off.

Kniphofia hybrids

Kniphofia Red-hot Poker

Well-established red-hot poker clumps are best divided in early spring. The divisions are often rather bare of roots but usually re-establish well. Many of the garden forms are hybrids that will not come true to type from seed. However, if you are keen to see what results—it could be something interesting—or would like to grow species, sow the seed in spring, uncovered, at around 68°F (20°C).

Koelreuteria bipinnata

Koelreuteria Golden-rain Tree

Golden rain trees can be raised from the seeds found in the papery seedpods. Sow them in spring at around 68°F (20°C). In addition, this is one of the few trees propagated by root cuttings, rather a tricky process for a tree. Little wonder that most examples are seedlings.

Kolkwitzia amabilis cultivar

Kolkwitzia Beautybush

This popular spring-flowering Chinese deciduous shrub is most often propagated by taking semi-ripe cuttings in summer. Winter hardwood cuttings will strike too and may be a better, though slower, method for those without mist.

Kunzea ericoides

Kunzea

Propagate these showy evergreen shrubs by sowing their very fine seed, which should be lightly mixed into the soil surface and kept at around 68°F (20°C), or by taking softwood or semi-ripe tip cuttings of non-flowering shoots during the warmer months. Late summer softwood cuttings seem to strike best but need mist to keep them from wilting.

Laburnum Golden Chain Tree

The species may be raised by sowing at 59–70°F (15–21°C) seed that has been soaked for 24 hours.

However, many garden plants are cultivars or hybrids that must be propagated vegetatively, either by taking hardwood winter cuttings or by grafting on to seedling stocks, usually of *L. anagyroides*.

Laburnum x *watereri* 'Vossi'

Lachenalia Cape Cowslip

Some species, such as the popular *L. bulbifera* and *L. aloides*, are very prolific, producing hundreds of bulbils around the parent bulb, but often these southern African bulbs are slow to produce offsets and are most commonly raised from seed. Sow in spring, lightly covered, at around 68–72°F (20–22°C). *L. bulbifera* can also be raised from leaf cuttings. Remove a leaf at flowering time, insert in potting mix and tiny bulbils will form around its cut base.

Lachenalia bulbifera

Lagerstroemia Crape-myrtle

These beautiful small trees may be grown from seed sown in spring at 68–77°F (20–25°C), but as most are cultivars or hybrids, vegetative methods are generally preferred. The most common is by striking semi-ripe cuttings in autumn. Hardwood cuttings taken during winter strike well if kept indoors with a little bottom heat, or in mild areas they can be kept outdoors.

Lagerstroemia indica 'S.D.H. Pluard'

Lagunaria Norfolk Island Hibiscus

Although this attractive evergreen tree can be propagated quite easily from seed, the seedpods contain irritant hairs that cause severe skin irritations. Consequently, if only a few plants are required cuttings are a more pleasant method to use. Softwood and semi-ripe cuttings taken during the warmer months strike fairly well, especially with mist and bottom heat.

Lagunaria patersonia

Lamium Dead Nettle

Known as dead nettle because their leaves resemble those of the stinging nettle but lack the stinging hairs, these quick-growing annuals and perennials often propagate themselves only too well, becoming rather invasive. Most garden plants are cultivars that must be propagated vegetatively. Divide established clumps from autumn to early spring or take cuttings of non-flowering stems during the warmer months.

Lamium galeobdolon 'Hermann's Pride'

Lantana

In the sub-tropics and tropics some lantanas are weeds. The ease with which they are propagated tells you why. Not only does the seed germinate rapidly at 68–77°F (20–25°C), softwood and semi-ripe cuttings will strike roots within two or three weeks if taken during the warmer months.

Lantana camara

Lapageria rosea 'Alba'

Larix decidua

Lathyrus vernus 'Cyaneus'

Laurus nobilis

Lavandula stoechas

Lavatera trimestris

Lapageria Chilean Bellflower

The Chilean bellflower grows well from seed sown in spring at around 68°F (20°C). The cultivars will not strike as cuttings and must be grown from layers, which accounts for the high price of the white-flowered form of *L. rosea*.

Larix Larch

With the exceptions of the hybrids and cultivars, larches are usually raised from seed, which should be sown in spring, preferably after having been stratified for about eight weeks. The seed germinates well at around 68°F (20°C) and should be lightly covered.

Although considerable work has been undertaken to grow larches easily from cuttings and tissue culture—because of their timber potential—most hybrids and cultivars are grafted onto young seedling stocks.

Lathyrus Sweet Pea, Wild Pea

Wild peas may be annual, like the sweet pea (*L. odoratus*), but many of them are perennials. Scarify the seed and soak for a day before sowing. While the perennials generally germinate at lower temperatures than the annuals, the difference is not critical. Clump-forming perennials, including the tuberous *L. tuberosus*, can be broken up in early spring, and semi-ripe cuttings taken in summer usually strike well.

Laurus Bay Laurel

The two laurel species are beautiful evergreen trees. The species may be raised from seed, although they and the cultivars are more commonly propagated by semi-ripe cuttings , 4–6 in (10–15 cm) long, taken in summer and autumn.

Lavandula Lavender

Lavenders germinate extremely freely from seed and can become weed-like in some areas. The selected forms must be propagated vegetatively. While old, established plants often self-layer or may be layered, the quickest production method is softwood cuttings, which can be taken at any time, with summer to early autumn being best. Mist and bottom heat help.

Lavatera Tree Mallow

The annual *L. trimestris* strains are easily raised from spring-sown seed. The perennials may also be raised from seed sown at around 68°F (20°C), and they also grow very readily from softwood and semi-ripe cuttings, which can be taken at any time, especially in the warmer months.

Ledum Labrador Tea

As with most erica family plants, these small shrubs produce capsules that contain many minute seeds. They germinate well if sown, uncovered, on finely sieved sphagnum moss, but it is usually simpler to take semi-ripe cuttings in summer and fall. Established plants often form natural layers that can be removed for growing on.

Ledum groenlandicum

Leonotis Lion's Ear

The commonly grown species in this genus are sub-shrubs. They sometimes form clumps of low growth that can be divided in winter and early spring, or more often it is possible to remove a few rooted shoots from the base of the foliage clump. Cuttings are also straightforward, semi-ripe cuttings striking well throughout the warmer months. Softwood cuttings strike, too, but really need mist to prevent wilting in summer.

Leonotis leonurus

Leontopodium Edelweiss

This well-known European alpine may be propagated by seed or division. The seed, which germinates more evenly if stratified for a few weeks, should not be covered and germinates best at around 68°F (20°C). Divide well-established clumps in late winter or early spring. Keep the divisions large so that they re-establish quickly.

Leontopodium alpinum

Leptospermum Tea Tree, Manuka

Cultivars of *L. scoparium* strike very readily from tiny cuttings of the tip growth. However, these may not be practical unless you have mist. Larger semi-ripe summer and autumn cuttings will strike, as will those of other species, but they will take longer. The species can be raised from seed sown, lightly covered, at around 68°F (20°C).

Leptospermum lanigerum

Leschenaultia

These beautiful small shrubs from Western Australia are propagated from seed or cuttings. The seed germinates at 64–75°F (18–24°C) but may take a considerable time. Soaking sometimes speeds germination. Softwood and semi-ripe cuttings strike variably during the warmer months. Some growers report success with winter root cuttings.

Leschenaultia biloba

Leucadendron and *Leucospermum*

Although these southern African shrubs differ markedly in their flowers, they are both proteaceous genera with similar requirements. Fresh seed germinates quite well if scarified or soaked before sowing. The young seedlings are, however, prone to collapse, usually due to fungal problems. Give them excellent ventilation and just enough water to keep them standing up. Semi-ripe cuttings strike best in the late summer and autumn but are unpredictable, often taking a long time for a low percentage strike.

Leucadendron 'Inca Gold'

Leucojum vernum

Leucothoe fontanesiana 'Rainbow'

Lewisia cotyledon

Liatris spicata

Libertia formosa

Libocedrus bidwillii

Leucojum Snowflake

Well known as late winter- and spring-flowering bulbs, there are also species that flower in late summer and autumn. All form clumps of bulbs and bulblets that can be broken up when dormant, though it is not recommended for division to be done regularly but only when the clumps lose vigor. Lightly covered seed sown in spring will germinate within a few weeks if kept at around 64°F (18°C).

Leucothoe

These mainly spring-flowering shrubs may be raised from seed or cuttings and sometimes by removing rooted suckers. The seed germinates best on a sphagnum-based mix, uncovered, at around 68°F (20°C), taking at least two weeks to show. Softwood and semi-ripe cuttings strike well from mid-spring to mid-autumn, especially with mist and bottom heat.

Lewisia

These rockery and alpine plants have a reputation for being difficult to cultivate, let alone propagate. However, most species, including the very popular *L. cotyledon*, are quite easily raised from spring-sown seed. The seed should be stratified for at least four weeks, then sown at 68°F (20°C), lightly covered. The young seedlings are inclined to damp off and good ventilation is important. Division is usually difficult, but the evergreen species sometimes form offsets that can be detached from the parent plant and grown on.

Liatris Gayfeather, Snakeroot

These North American perennials are easily propagated from seed, division or by treating the young shoots as cuttings. The seed is best sown uncovered and germinates at around 68°F (20°C). Divide large clumps in winter, replanting the knobbly rootstock so the top is level with the soil surface. Softwood cuttings of non-flowering shoots strike well, especially under mist.

Libertia

These fibrous-rooted perennial iris relatives are very easy to propagate. Established clumps withstand division at almost any time and even small pieces will quickly recover and grow. Seed sown in spring readily germinates at around 64°F (18°C) and soon develops into small foliage clumps. Some species send out runners that can be removed and grown on.

Libocedrus

These sub-tropical and warm-temperate conifers may be raised from seed or cuttings. Seeds of species from New Caledonia do not require stratifying but need more warmth than those of the South American and New Zealand species. Semi-ripe spring to autumn cuttings, preferably with a small heel, often take a while to strike but are fairly reliable.

178

Ligularia and *Farfugium* Leopard Plant
Most species of these very hardy perennials are easily raised from seed, but because established clumps divide so readily through winter and into early spring, division is the preferred method for species and cultivars.

Ligularia przewalskii

Ligustrum Privet
Privets grow easily from seed but because so many garden plants are selected forms, cuttings are generally preferred. Softwood and semi-ripe cuttings strike quickly and well during the warmer months and the deciduous types can also be grown from winter hardwood cuttings. It is often possible to remove rooted basal suckers.

Ligustrum obtusifolium

Lilium Lily
Lilies are the archetypal scaly bulbs. Most are easily propagated by breaking up the bulbs into individual scales in late autumn, keeping them in moist potting mix and waiting for the young plants to develop. Many also produce bulbils around the main bulb during the growing season or can be induced to form bulbils in the leaf axils, if that does not occur naturally.

The seed germinates well, but remember that many garden lilies are hybrids, so seed is really only suitable for propagating species or producing new hybrids. The seed, which sometimes needs stratifying, should be sown covered at around 68°F (20°C). The young seedlings are tender and are often best left in the seed trays until the foliage dies back in autumn.

Lilium 'Compass'

Linaria Toadflax
These annuals, biennials or short-lived perennials are usually increased from seed, which may be sown in autumn or spring. The seed germinates quickly even at cool temperatures and needs only a light covering. The perennials may be divided in winter when dormant.

Linaria purpurea

Lindera Spice Bush
Known for their extreme hardiness and aromatic foliage, these shrubs produce small globular fruits that contain a single seed. Sow the seed fresh in fall outdoors (having first removed any fruit pulp), or stratify it for six weeks before sowing indoors to simulate an early spring.

Semi-ripe cuttings taken in summer and fall strike reasonably well but need to be well-established before winter if they are to survive harsh cold outdoors.

Linnaea Twinflower
This dwarf, creeping shrub is most often propagated by removing small rooted pieces in winter. Softwood and semi-ripe cuttings will also strike but need to be kept cool and moist.

Lindera obtusiloba

Liquidambar orientalis

Liriodendron tulipifera

Lithodora diffusa

Lobelia erinus

Lomatia ilicifolia

Lonicera sempervirens 'Sulphurea'

Liquidambar Sweetgum

Although new plants can be raised from seed, the best foliaged forms should be perpetuated by taking cuttings or by budding on to seedling stocks. Stratify the seed for eight to 12 weeks before sowing in spring at around 68°F (20°C).

Semi-ripe summer cuttings and winter hardwood cuttings strike equally well, though summer cuttings need to be well established by winter to survive outdoors.

Liriodendron Tulip Tree

Except for the variegated foliage and other fancy forms, which are grafted, most plants are raised from seed. Sow the seed outdoors in autumn or stratify in a refrigerator for eight to 12 weeks before sowing in spring. Lightly cover the seed and keep it at around 68°F (20°C).

Lithodora

Still widely known as *Lithospermum*, these popular, usually blue-flowered, carpeting plants are easily propagated by softwood or semi-ripe tip cuttings taken from late spring until autumn. Despite being trailers, they do not readily form natural layers, but can be encouraged to do so by mounding soil over the stems.

Lobelia

The annual bedding lobelias grow easily from seed sown in late winter and spring at 70°F (21°C), as do the perennial species, but selected forms must be propagated vegetatively. Established clumps can be broken up in early spring, and most will strike from cuttings of new fleshy shoots throughout the growing season.

Lomatia

Found in South America and Australia, these protea relatives are evergreen shrubs and small trees that may be raised from seed or cuttings. The seed should be covered and germinates at fairly cool temperatures, around 64°F (18°C), but it may take some time to show. Semi-ripe cuttings taken in summer and autumn strike moderately well, with mist and mild bottom heat helping considerably.

Lonicera Honeysuckle

Honeysuckles have a well-deserved reputation for vigor. The seed of the species will germinate and develop quickly, but most garden plants are raised from cuttings or layers. Softwood and semi-ripe cuttings can be taken from spring to autumn and hardwood cuttings in late autumn and winter. Any long, arching stems can be pegged down as layers.

Loropetalum Fringe Flower

The flowers of the two species in this genus clearly show that the plants are related to the witch hazels (*Hamamelis*). The commonly grown species, *L. chinense*, is usually propagated by taking softwood or semi-ripe cuttings in summer or autumn. The young plants are rather frost tender and should be kept under cover for their first winter. Spring sown seed germinates well but the seedlings vary considerably in flower quality.

Loropetalum chinense cultivar

Luculia

These beautiful winter-blooming, fragrant-flowered, evergreen shrubs from the Himalayan region are easily raised from fresh seed. Sow at around 68°F (20°C), lightly cover, and it should germinate within three weeks. The seedlings often flower in their second year. Propagate the cultivars by taking semi-ripe cuttings in late summer or autumn.

Luculia gratissima 'Early Dawn'

Lupinus Lupin, Lupine

As befits plants that can become rather invasive, lupin seed germinates very readily. Seed is the best method for propagating the species and the only method for annuals such as Texas bluebonnet (*L. texensis*). Soak or scarify it before sowing, lightly cover the seed and give it a temperature of around 64°F (18°C).

Established clumps can be divided in late winter or early spring and the strong new shoots can be used as softwood cuttings. The shrubbier types are most easily propagated by softwood or semi-ripe cuttings during summer and autumn.

Russell lupins

Lychnis Campion, Catchfly

Although most *Lychnis* are perennials, many of them are treated as annuals and raised afresh from seed each year. Sow the seed uncovered at around 68°F (20°C). The more genuinely perennial types left in the ground over winter can be also raised from seed or divided in late winter to early spring.

Lycoris Spider Lily

Propagate these bulbs by breaking up established clumps in summer when they are dormant or by raising young plants from seed. Stratification is not essential but will result in more complete and even germination. Sow seed lightly covered at around 68°F (20°C); germination will begin in 2–3 weeks.

Lychnis flos-jovis

Lysichiton Skunk Cabbage

These hardy rhizomatous perennials may be divided when dormant or raised from seed. Divide near the end of the dormant period, in late winter or early spring, and keep the divisions large so they re-establish quickly. The seed germinates in spring at cool temperatures, around 59°F (15°C), and needs only a light covering.

Lysichiton americanus

Lysimachia clethroides

Lysimachia Loosestrife

Loosestrifes are very easily propagated by cuttings of the vigorous young shoots or by dividing established clumps in late winter or early spring. The spreading forms, such as the well-known creeping Jenny (*Lysimachia nummularia*), often strike roots as they spread, allowing small rooted pieces to be broken off and grown on. Because they so readily propagate by other means, seed is rarely used, though it usually germinates without any difficulty.

Lythrum salicaria 'Rose Queen'

Lythrum Loosestrife

These annuals, perennials and small shrubs are very easily propagated, often self-sowing. The seed germinates throughout the warmer months and will appear within two to three weeks of sowing at 68°F (20°C). The perennials can also be divided when dormant in winter and very early spring and their young non-flowering shoots can also be used as cuttings during the growing season.

Macadamia integrifolia

Macadamia

The hardness of macadamia nuts is well known, and the nuts, of course, enclose the seeds. Consequently, softening the seeds is an important part of germinating macadamia nuts. Scarify then soak for 48 hours before sowing, well covered, at around 72°F (22°C). The best fruiting forms are usually propagated vegetatively, sometimes by taking semi-ripe summer cuttings, but more often by grafting on to seedling stocks.

Macleaya cordata

Macfadyena Cat's Claw Vine

These vigorous climbers, sometimes invasive in the tropics, are attractive plants where they can be controlled. Propagation is usually by taking cuttings or layers. Softwood and semi-ripe cuttings taken throughout the warmer months strike freely, especially with mist and bottom heat. Layers pegged down in spring should be well struck by the following growing season.

Macleaya Plume Poppy

While these vigorous perennials can be increased from seeds it is generally far easier to divide an established clump in early spring as it starts into growth. Alternatively, winter root cuttings are a reliable way to produce a large number of plants.

Magnolia 'Mark Jury'

Magnolia and *Michelia*

These beautiful shrubs and trees are propagated in several ways. The species may be grown from fresh seed sown at 70°F (21°C), with the deciduous types often requiring stratification before sowing.

The hybrids and cultivars can be grown from summer cuttings, but it is important that deciduous cuttings are well established before winter or they need to be over-wintered indoors. Mist and bottom heat are almost essential for success.

Layering and aerial layering are successful and although they take a considerable time, their reliability makes the wait worthwhile, especially if you do not have sophisticated propagation equipment.

Magnolia stellata

Mahonia

Stratified *Mahonia* seed germinates in about four weeks if sown in spring at around 68°F (20°C). It is usually easier, however, to take cuttings, and that way you know exactly what quality of flowering and fruit to expect. Semi-ripe cuttings taken in summer and autumn are usually well established by winter. Take care handling those prickly leaves.

Mahonia lomarifolia

Malus Apple, Crabapple

While true species can be propagated from seed, it is necessary to bud or graft named varieties, be they ornamental crabapples or edible fruiting cultivars. The seed, either for species trees or grafting stocks, should be stratified for up to 12 weeks before sowing in spring at around 68°F (20°C). Some cultivars and species may be grown on their own roots by taking winter hardwood cuttings but most often scions of known varieties are budded or grafted on to seedling stocks or established plants.

Malus halliana

Malva Mallow

Mallows are very easy to propagate, as you would expect of plants that are inclined to be rather invasive. The seed is usually up within two weeks if sown at 68°F (20°C), and softwood or semi-ripe cuttings strike almost as quickly if taken during the growing season.

Malva sylvestris

Mammillaria Pincushion Cactus

Mammillarias are clustering cacti and the offsets produced around the main stem are quite easy to separate and grow on. These little cacti usually flower prolifically and develop conspicuous soft fruits. Remove the seeds by hand or steep in water; they will germinate fairly quickly if sown at around 68°F (20°C). To prevent damping off, the young seedlings need perfect ventilation and just enough moisture to keep them turgid.

Mammillaria backebergiana

Mandevilla

This genus of climbers is notable for its showy flowers and the large bean-like seedpods that follow. The seeds within the pods germinate well if kept warm (around 72°F /22°C) and even better if soaked before they are sown. Semi-ripe cuttings taken at any time during the warmer months strike well, but they need to be well established before winter or they may collapse with the colder weather.

Mandevilla boliviensis

183

Maytenus boaria

Mazus radicans

Meconopsis cambrica

Melaleuca ericifolia

Melia azederach

Maytenus Mayten

This quite large Central and South American genus is represented in cultivation largely by just one tree, the mayten (*M. boaria*), which has an appearance rather like a weeping willow, but evergreen. The finest weeping forms are best propagated vegetatively by taking softwood and semi-ripe cuttings during the growing season. They strike quite freely, more so with mist and mild bottom heat.

Mazus

These small, spreading perennials are very easily divided, having stems that form roots where they touch the ground. The clumps can be broken up at any time but re-establish most readily in early spring. Alternatively, the stem tips can be used as cuttings and will strike very quickly.

Meconopsis

This genus of poppies is famed for its blue-flowered Himalayan species, which are the most difficult in the genus to cultivate. The seed of most species, especially Welsh poppy (*M. cambrica*), germinate readily if fresh, but unless they are kept cool and moist with high humidity the Himalayan seedlings are inclined to collapse. Sow the seed, lightly covered, on a humus-rich mix and keep it very moist and cool until the seedlings are ready to pot. Well-established clumps growing under ideal conditions can be divided, but I am inclined to think that if you have a good plant then leave it alone.

Melaleuca Paperbark

As many a Florida resident can testify, paperbarks grow well from seed. If lightly covered and kept at around 68°F (20°C), the seed of most species will have germinated within three weeks of sowing. Softwood and semi-ripe cuttings taken during the growing season are the preferred propagation method for the hybrids and cultivars. If you have mist, try using very small tip cuttings—they should strike particularly quickly.

Melia Bead Tree

While these evergreen and deciduous trees often set good crops of seed, because the garden specimens tend to be selected forms they are usually propagated vegetatively. This is done by taking heeled cuttings of non-flowering sideshoots in late summer or by root cuttings taken in late autumn and early winter.

Menziesia Minnie Bush, Fool's Huckleberry

These deciduous shrubs, ericaceous plants from North America and Asia, may be propagated by seed or cuttings and the low-growing species sometimes self-layer. The seed, which is very fine, is usually best stratified for 6–8 weeks before sowing on finely sieved sphagnum moss. It should not be covered but should be shaded and kept cool and moist. Softwood and semi-ripe spring and summer cuttings strike well, appreciating mist and mild bottom heat.

Meryta sinclairii

Meryta Puka
These large-leaved shrubs and trees from the southern Pacific are easily propagated by taking tip cuttings in summer and autumn but the size of the stems and foliage can make the cuttings very unwieldy. Consequently, seed is sometimes an easier choice. Sow, covered, at around 72°F (22°C). Cultivars, such as the variegated forms of *M. sinclairii*, must be vegetatively propagated.

Mesembryanthemum and other ice plants
Ice plants are low, spreading succulents capable of carpeting large areas but they do not always strike roots as they spread. Those that do are easily propagated by removing small rooted pieces for growing on, otherwise you will need to strike softwood or semi-ripe cuttings during the warmer months or sow the seed, uncovered, at around 68°F (20°C).

Mesembryanthemum crystallinum hybrids

Mespilus Medlar
Like the quince, the medlar was once very popular for its fruit, which was widely used in jams, jellies and preserves. Now uncommon, it is also worth growing for its fragrant white flowers and russet fall foliage. The best forms are grafted onto seedling stocks that are raised by sowing stratified seed in spring at around 64°F (18°C). The seed should be covered.

Metasequoia Dawn Redwood
This ancient tree, a deciduous conifer, can be raised from seed or cuttings. Chill the seed and sow in the open in autumn to germinate in spring or stratify in a refrigerator for 12 weeks prior to spring sowing at 59–72°F (15–22°C) with a light soil covering. Semi-ripe cuttings under mist strike well in summer and early winter hardwood cuttings kept under cover with mild bottom heat usually have roots by spring.

Mespilus germanica

Metasequoia glyptostroboides

Metrosideros Pohutukawa, Rata
The New Zealand pohutukawa (*M. excelsa*) and its tropical relatives are spectacular summer-flowering shrubs and trees. They are easily raised from seed sown at around 68°F (20°C), but as many of the garden plants are hybrids or cultivars, cuttings are the more common propagation method. Use softwood or semi-ripe cuttings from spring to autumn and be prepared to wait a little while unless you have access to mist and bottom heat. Some species behave differently depending on how they are propagated. For example, *M. carminea* grows as a climber from seed, but cuttings from adult plants have a shrubby growth habit.

Metrosideros excelsa

185

Mimulus guttatus

Mirabilis jalapa

Mitraria coccinea

Monarda 'Fire Beacon'

Monstera deliciosa

Morus nigra

Mimulus Musk, Monkey Flower

The musks are well known as bedding annuals and perennials and also include a few evergreen shrubby species. The annual strains are raised from seed that is sown, uncovered, at around 68°F (20°C). Divide the perennials in late winter once well established or propagate by softwood cuttings taken from spring to autumn. The shrubby species, of which *M. aurantiacus* (syn *M. glutinosus*) is the most widely grown, are struck from softwood or semi-ripe cuttings in summer and early autumn.

Mirabilis Umbrella Wort, Marvel of Peru

These tuberous perennials are easily broken up when dormant; larger numbers can be raised from seed. The seed of North American species may benefit from stratification but otherwise simply sow it, lightly covered, indoors in late winter to early spring at around 68°F (20°C). By the time it is warm enough outdoors the seedlings should be ready for planting out.

Mitraria

The sole species in this Chilean genus is a rather straggling, semi-climbing, evergreen shrub grown for its striking orange flowers. Softwood and semi-ripe cuttings strike freely throughout the growing season and layers may be pegged down at any time.

Monarda Bergamot

The perennial bergamots form large clumps that divide readily from late autumn to early spring, preferably in late winter just as they start into growth. Cultivars can also be propagated by taking cuttings of the vigorous basal shoots. The species may be raised by sowing the seed, lightly covered, at around 68°F (20°C).

Monstera and *Philodendron* Fruit Salad Plant

Popular as houseplants in temperate zones, these vigorous climbers can cover an enormous area when grown in the tropics and often develop an impressive array of aerial roots. When used as houseplants their fruits are rarely seen, but in tropical gardens the arum-like flowers are followed by large, edible, pulpy fruits. If sown at around 79°F (26°C), the seed germinates within two to three weeks. Semi-ripe cuttings, although often large and unwieldy, strike quickly in summer and establish well.

Morus Mulberry

Cuttings are usually the first choice for propagating mulberries, either semi-ripe summer cuttings or winter hardwood, but because these can sometimes be reluctant to strike the best forms are sometimes grafted on to seedling stocks. Seedlings for grafting can be raised from spring-sown seed kept at around 72°F (22°C).

Murraya

These citrus relatives, most commonly represented in cultivation by orange jessamine (*M. paniculata*), are usually raised from cuttings. Softwood and semi-ripe cuttings strike well during the warmer months, especially with mist and mild bottom heat. The brightly colored berries contain seeds that germinate well but cuttings yield flowering plants immediately.

Murraya paniculata

Musa Banana

Commercial bananas are sterile, seedless hybrids that must be vegetatively propagated. This is usually done by removing rooted offsets from the base of established plants during spring and summer. The true species produce fertile fruit—bananas with seeds—and may be propagated by removing the seeds from the pulp then sowing them, lightly covered, at around 75°F (24°C).

Musa x paradisiaca

Muscari Grape Hyacinth

Known for their ability to quickly form large clumps, these small, spring-flowering bulbs are most readily propagated by division after the foliage has died back. The species may also be raised from seed: sow, lightly covered, outdoors in autumn when fresh or in spring, possibly after brief stratification.

Muscari neglectum

Myoporum Boobialla, Ngaio

The evergreen shrubs and trees of this southern hemisphere genus are easily propagated by seed or cuttings. Most species produce purplish drupes containing seeds that germinate quickly at around 68°F (20°C). Softwood and semi-ripe cuttings strike very well at any time during the growing season or year-round in mild climates.

Myoporum laetum

Myosotidium Chatham Islands Forget-me-Not

This large-leafed and impressively flowered perennial is usually raised from seed, which germinates freely if sown, covered, in humus-rich soil and kept cool. The seeds continue to germinate over quite a long period, with odd seedlings continuing to appear well after the first ones show. Established clumps can be divided with care but, considering how easy it is to raise seedlings, it is inadvisable to disturb good plants.

Myosotidium hortensia 'Alba'

Myosotis Forget-me-Not

The common forget-me-not (*M. sylvatica*) and its bedding plant strains pop up everywhere once you let your original plants run to seed. Less vigorous species do not always germinate as well but if sown fresh and uncovered at around 64–68°F (18–20°C), most produce a good crop of seedlings. Clump-forming perennial forget-me-nots can be divided, preferably in late winter so that they can start growing right away. Some also have shoots long enough to be used as softwood cuttings.

Myosotis sylvatica

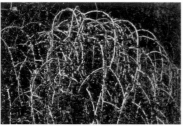

Myrsine divaricata

Myrtus communis var *italica*

Nandina domestica

Narcissus 'February Gold'

Nelumbo nucifera

Myrsine

The five species in this genus have a widely scattered and interesting distribution that encompasses China, the Azores and New Zealand among other places. Most species are propagated from cuttings, either softwood or semi-ripe, taken from late spring to early autumn.

The seed from the small, single-seeded fruit may also be germinated but considering the ease of taking cuttings it is seldom worth the effort.

Myrtus Myrtle

This once large genus has been extensively revised and now includes just two species, of which the common myrtle (*M. communis*), in all its varieties and cultivars, is the better known.

Although the basic species can be raised from seed sown in spring at around 68°F (20°C), most garden varieties do not come true to type from seed and are propagated by softwood and semi-ripe cuttings taken from late spring to early autumn.

Nandina Heavenly Bamboo

Heavenly bamboo forms clumps but the woodiness of its stems makes it difficult to divide. Instead, try taking cuttings of the soft growth at the stem tips. This usually strikes well in summer and autumn and quickly produces neat, compact plants. The roots, however, are brittle so take care when transplanting the cuttings.

Narcissus Daffodil

Daffodils are the archetypal spring-flowering bulbs. In many areas they naturalize well, multiply freely and can be lifted and broken up when dormant. Commercial growers speed up the natural rate of increase by using a rather complicated technique called double or twin scaling. Consult specialist literature for the details of this method. Use powdered insecticides to coat stored bulbs if narcissus fly is a problem in your area. Of course, the species can be raised from seed, which must be fresh. Usually it will be at least the third season before you see any good flowers.

Nelumbo Sacred Lotus

Lotuses are increased by seed or by dividing the rhizomes when dormant. Gently scarify seeds before and sow in water at around 72°F (22°C). The seeds that float are infertile and should be disposed of. Once the seeds sprout, place them in individual pots of loamy soil, put a layer of gravel on the soil to keep it in the pot, then immerse so that the soil surface has a 2 in (5 cm) covering of water. Keep the seedlings in a bright place and increase the pot size and water depth as they grow.

Nemesia

Nemesias may be annual or perennial. The annuals tend to be very short-lived and must be replaced every three months or so for a continuous display. It is fortunate then that the seed sprouts like grass in just over a week if sown at around 68°F (20°C). The perennials are just as easy from seed but the cultivars must be propagated vegetatively. Take softwood cuttings from late spring until mid-autumn.

Nemesia hybrids

Nepeta Catnip, Catmint

While cultivars such as 'Six Hill's Giant' are grown from small softwood cuttings, most catnips and catmints are raised from seed, even though they are perennials. That is because the seed germinates so quickly and reliably. Sow it at around 68°F (20°C), cover it lightly and it should be up in a week or so.

Nepeta x *faassenii*

Nerine Spider Lily

The usual method of propagating nerines is to just leave them to get on with it. The bulbs multiply rapidly and soon form clumps that are large enough to be broken up immediately after the foliage has died away. The seed germinates well if sown at around 68°F (20°C) and lightly covered.

Nerine bowdenii

Nerium Oleander

Although seldom raised from seed, oleanders germinate well. Instead, they are usually grown from semi-hardwood cuttings taken in late summer, which are reliable but slow to root, taking up to 12 weeks, even with bottom heat. If you do sow seed, say for raising your own hybrids, keep it at around 75°F (24°C) and be very wary of damping off.

Nerium oleander 'Luteum Plenum'

Nicotiana Tobacco

Tobaccos may be annual or perennial and some of the perennials are treated as annuals. Sow seed, uncovered, at around 72°F (22°C). The truly perennial types can be raised by taking spring cuttings of the strong basal shoots or by slicing off shoots with roots attached. Divide very large clumps in early spring. The few shrubs in the genus are most easily propagated by softwood or semi-ripe summer cuttings.

Nicotiana sylvestris

Nierembergia Cup Flower

This South American genus includes annuals, perennials and sub-shrubs that may be raised from seeds, cuttings or divisions. The seeds of the annuals germinate quickly at around 72°F (22°C); that of the perennials taking a little longer. Clump-forming types can be divided in late winter or early spring as they start into growth. During the growing season the sub-shrubs and most of the perennials can also be raised from small tip cuttings.

Nierembergia repens

Nothofagus sp.

Nymphaea hybrids

Nyssa sylvatica

Ochna serrulata

Oenothera biennis

Olea europaea

Nothofagus Southern Beech

These trees, dominant in many of the forests of New Zealand and southern South America, are usually raised from seed. The seed of the few deciduous South American species definitely germinates more evenly if stratified for eight weeks before sowing in the spring, and stratification would also probably be beneficial for most other species. Lightly cover the seed and keep at 59–68°F (15–20°C).

Nymphaea Waterlily

Waterlilies develop from tubers, rhizomes or stolons, depending on the species. Simply divide or cut up the rootstock when it is at its most dormant and replant the divisions as the weather warms.

Nyssa Tupelo

Known for their ability to grow in wet soil and for their autumn foliage color, this genus of five species of deciduous trees is propagated in various ways, including by seed. Sow seed outdoors when ripe or stratify then sow in spring. Alternatively, raise from late summer cuttings, or layers, or by grafting, usually on to seedlings of *N. sylvatica*. Summer cuttings need to be well established before winter or they may require protection from severe cold.

Ochna Bird's Eye Bush, Mickey Mouse Plant

Seed is the most common propagation method for this tropical genus of shrubs and trees. The fruit of the commonly grown species is very showy and contains hard seeds. Soak them for 24 hours before sowing, covered, at around 72°F (22°C). Semi-ripe cuttings strike quite well but with such a proliferation of quick-germinating seeds they are seldom used.

Oenothera Evening Primrose

Evening primroses may be biennial or perennial. The biennials are raised from seed sown, uncovered, at around 72°F (22°C). The perennials may also be propagated by division, if they form large clumps, or by taking cuttings of the vigorous basal shoots as they develop in spring. Winter root cuttings are probably the fastest way of building up large numbers of selected plants.

Olea Olive

Although grown commercially for its fruit, the olive is also a popular ornamental tree. Olive seeds will germinate if soaked and scarified but as most of the garden forms are cultivars, vegetative propagation is preferred. Softwood or semi-ripe summer cuttings that have been wounded will strike, but may take up to four months to do so. Cuttings of harder wood taken later also strike but take even longer. Olives are sometimes grafted.

Olearia Daisy Bush

The tree daisies are usually propagated by cuttings. They strike well from softwood or semi-ripe cuttings taken from late spring until mid-autumn. The seed, which is often produced in abundance, germinates well, though this method is only suitable for the species. It should be as fresh as possible and sown lightly covered at 59–64°F (15–18°C).

Olearia macrodonta

Ophiopogon and Liriope Mondo Grass, Lilyturf

One look at these grassy, clump-forming relatives of the lilies tells you they can be divided. Do it in late winter or very early spring.

Seed propagation differs between the two genera: *Ophiopogon* usually benefits from a brief period of stratification, and sow covered, at around 64°F (18°C); Soak *Liriope* seed before sowing, covered, at around 68°F (20°C).

Ophiopogon planiscapus 'Nigrescens'

Ornithogalum Chincherinchee, Star of Bethlem

These southern African and Mediterranean bulbs are very easily propagated by dividing established clumps when dormant, usually in late winter or very early spring. They also set plenty of seed: sow in spring, lightly covered, at around 68°F (20°C).

Ornithogalum thyrsoides

Osmanthus

Found almost exclusively in temperate East Asia, these evergreen shrubs are most often propagated by taking semi-ripe cuttings in summer and early autumn. They can also be grown from softwood cuttings kept under mist or by taking larger hardwood cuttings in winter.

Osmanthus x burkwoodii

Ostrya Hop Hornbeam

Resembling its relative the hornbeam (*Carpinus*) in foliage, these deciduous trees have fruiting catkins that enclose a single seed within an arrangement of overlapping bracts. This seed is easily collected and is the main propagation method. It can be sown outdoors in fall or stratified for at least six weeks before being sown, lightly covered, in spring at around 68°F (20°C).

Ostrya carpinifolia

Pachysandra

The common *P. terminalis* is a creeping sub-shrub that strikes roots as it spreads. If you simply want a few more plants, it is often possible to break off rooted pieces and grow them on. If greater numbers are required, softwood and semi-ripe cuttings strike well at any time.

The less common *P. procumbens* is a perennial with less woody stems and is most easily propagated by division in late winter and early spring.

Pachysandra terminalis

Pachystachys lutea

Pachystachys Golden Candles

Grown as houseplants in the temperate areas or outdoors in tropical gardens, these perennials and shrubs produce flower heads with showy bracts. Seedheads follow but as most of the cultivated forms are not species, vegetative propagation is preferred to seed-raising. Given warmth and humidity, softwood and semi-ripe cuttings strike freely during the warmer months.

Pachystegia insignis

Pachystegia Marlborough Rock Daisy

This low shrub from the windswept coast of the northeastern South Island of New Zealand is grown as much for its large, heavily felted, leathery leaves as for its show of white daisies. Fluffy seedheads follow the flowers and contain many small seeds that germinate well if sown, lightly covered, at around 64°F (18°C).

Softwood and semi-ripe cuttings usually strike within a few weeks, especially under mist with mild bottom heat.

Paeonia hybrids

Paeonia Peony

Peony seed germinates well if sown in spring at a temperature of around 70°F (21°C), but may require up to ten weeks of stratification and can take up to two months to germinate. Hybrids of perennial species are usually grown from divisions.

Tree peonies such as *P. delavayii* and *P. lutea* are often raised from seed. *P. suffruticosa* cultivars can be grown from root cuttings, but as these are rather unreliable, they are more commonly grafted on to seedling rootstocks. Layering is a reliable method with all types of tree peonies, but it is a very slow way to increase the stock.

Butia capitata

Palms

Although it may be possible to use tissue culture, palms are nearly always propagated from seed, except in the odd cases where basal suckers appear. Palm seed varies greatly in its requirements and is something of a specialized field. *Phoenix canariensis* germinates easily, as do a few others, but the seeds of some have such hard coats that there are difficulties with pre-sowing treatment. It is often better to avoid the worries and the wait and spend the extra to buy pre-sprouted seed.

Pandorea pandorana

Pandorea

These vigorous climbers produce conspicuous seedpods and can be raised from the seeds within, but because they layer freely and strike well from cuttings, they are usually vegetatively propagated. Softwood and semi-ripe cuttings strike fairly quickly at any time during the growing season. If you want to try the seed, sow it, covered, at around 68°F (20°C).

192

Papaver Poppy

Poppies are known for seeding heavily and coming up everywhere. So not surprisingly seed is the main propagation method. The requirements vary a little with the species but sowing the seed, uncovered, at around 68°F (20°C) is suitable for most of them. Those that form clumps, most notably *P. orientale*, can be divided when dormant.

Papaver nudicaule

Parahebe

These small sub-shrubs, mainly New Zealand natives, often self-layer as they grow and propagation is simply a matter of breaking off rooted pieces for growing on. Softwood cuttings strike quickly at any time, especially during the warmer months. The seed germinates at relatively cool temperatures and should be left exposed to light.

Parahebe lyalli

Parrotia Persian Witch Hazel, Persian ironwood

The sole species in this genus is a deciduous tree that bears distinctive deep red flowers in spring before its foliage appears. It also has attractive flaking bark and colorful autumn foliage. Seed is the usual propagation method. Sow outdoors when ripe, or stratify for eight weeks before sowing in spring at around 68°F (20°C). This tree can also be layered, usually aerially, because it has few branches near the ground.

Parrotia persica

Parthenocissus Boston Ivy, Virginia Creeper

Plants capable of covering the area that these can are usually easy to propagate and *Parthenocissus* is no exception. Semi-ripe cuttings taken during the growing season are the fastest way to produce new plants. Winter hardwood cuttings also strike well.

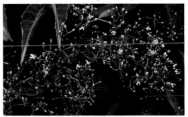
Parthenocissus henryana

Passiflora Passionflower

Passionfruits are full of seeds, as anyone with even passing familiarity with desserts knows. Those hard black seeds germinate best at around 75°F (24°C) and should be only lightly covered. The vines will also grow from softwood and semi-ripe cuttings taken from late spring to mid-autumn, and this is a quicker way to produce flowering plants.

Passiflora coccineus

Paulownia Princess Tree, Foxglove Tree

Paulownias grow extremely rapidly after leafing up in the spring and cuttings can be taken as soon as the new growth has hardened slightly. The struck cuttings are best over-wintered indoors or they may collapse with severe frosts. Root cuttings, not easy things to take from established trees, will strike in winter. The seed rattles noisily in its pods. Sow it fresh and uncovered at around 64°F (18°C). If the seed has been stored, stratify it for a few weeks before sowing.

Paulownia tomentosa

193

Pelargonium cordifolium

Penstemon pinifolius

Pentas lanceolata cultivars

Perovskia atriplicifolia

Persea americana 'Hass'

Persicaria campanulata

Pelargonium Geranium

The optimum conditions vary, but the seed of most pelargoniums germinates well at 61–77°F (16–25°C) and often benefits from scarification before sowing. Many of the garden plants are hybrids or cultivars and these are easily propagated by softwood or semi-ripe cuttings taken at any time during the growing season.

Penstemon

Species penstemons can be raised from seed. Being sown, covered, at around 64°F (18°C) suits most of them. The fancy hybrids, however, must be vegetatively propagated and many growers prefer to do the same with the species. Large clumps can be divided in late winter and early spring but many penstemons do not form clumps suitable for division. Consequently, taking cuttings of the vigorous young shoots is the preferred method.

Pentas Star Cluster

This tropical genus includes perennials and biennials that are often treated as annuals. The biennials are raised from seed, which germinates quickly if sown in spring, uncovered, at around 72°F (22°C). The perennials may also be raised from seed but are more commonly propagated by taking softwood and semi-ripe cuttings during the warmer months.

Perovskia Russian Sage

This genus of sub-shrubs is usually grown in the perennial border and may be increased by seed or cuttings. Spring-sown seed germinates at around 68°F (20°C) and requires no special treatment. Softwood and semi-ripe cuttings of non-flowering shoots taken in spring and summer strike readily but need to be well established before winter.

Persea Avocado

It is a common plant propagation lesson for children to sprout an avocado seed by suspending it on toothpicks on the edge of a glass or jar of water. If the base of the seed stays in contact with the water it will sprout. Of course, the seed can also be germinated in soil, partly covered, at around 77°F (25°C). Also, it is possible to strike semi-ripe cuttings in summer and early autumn. The best fruiting forms, however, are usually grafted on to seedling rootstocks in winter.

Persicaria Knotweed

Now separated from the once large genus *Polygonum*, this genus is represented in cultivation by several of its perennial species. These can be raised from seed but are so easily divided that it is seldom necessary. Rooted pieces can be removed at any time and grown on or well-established clumps may be broken up completely in early spring as the new growth commences.

Petunia

Until the advent of the new, more reliably perennial petunias, almost all stocks were raised from seed. The seed needs light and warmth to germinate—give it around 75°F (24°C) and do not cover it. The more compact, hardier forms can be propagated by softwood cuttings at any time during the growing season and will sometimes self-layer.

Petunia cultivars

Phellodendron Cork Tree

Known for the corky bark of some species, these deciduous trees produce berry-like fruits containing seeds that can be sown outdoors in fall or in spring after stratification. Sown covered, at 59–68°F (15–20°C), it will start to germinate in about three weeks. The seedlings can be grown on or used as grafting stocks for selected forms. Plants can also be raised from summer semi-ripe cuttings or winter hardwood cuttings.

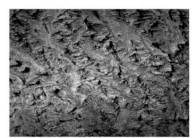

Phellodendron amurense

Philadelphus Mock Orange

These fragrant-flowered deciduous shrubs are usually propagated by cuttings: semi-ripe in summer with bottom heat and mist or winter hardwood cuttings kept outdoors. Layering also works well but is a slow method compared to cuttings. Although hybrids, such as *P. + virginalis*, should be propagated vegetatively, the species can be raised from seed. Sow it in spring with cool conditions—around 59°F (15°C)—after first stratifying for four to six weeks.

Philadelphus 'Enchantment'

Phlomis

This genus of perennials and sub-shrubs with felted leaves is propagated by seed, division or cuttings. The seed should be sown in spring, covered, and germinates at moderate temperatures of around 64°F (18°C). The clump-forming types can be divided in late winter or early spring after the worst frosts, while the woodier species are best propagated by taking warm season cuttings of non-flowering stems.

Phlomis fruticosa

Phlox

Phlox propagation varies depending on the plant type. The annual strains of *P. drummondii* germinate well at around 64°F (18°C) if lightly covered, while many of the perennial species need to be stratified briefly and germinate better with slightly more warmth—around 68°F (20°C). The hybrids and cultivars of the rock phloxes and the paniculata types can often be divided in late winter. Also, *P. paniculata* cultivars can be propagated by root cuttings. Softwood and semi-ripe cuttings taken from late spring to early autumn are preferred for the rock phloxes and other trailing types.

Phlox paniculata 'Pink Gown'

Phormium tenax

Photinia x *fraseri* 'Red Robin'

Phyllodoce x *intermedia* cultivar

Physalis alkekengi

Physostegia virginiana

Phyteuma spicatum

Phormium New Zealand Flax

These vigorous evergreen perennials can be raised in large numbers from seed but most garden examples are cultivars that must be propagated vegetatively to retain their leaf color. Late winter divisions of well-established clumps will recover quickly provided you do not make the divisions too small.

If you want to try the seed—bronze forms are often true to type—sow it at around 64°F (18°C) and cover lightly.

Photinia

Many of the common garden photinias are hybrids or cultivars and are generally grown from semi-ripe cuttings taken in summer or autumn. The species may be grown from seed sown in spring at around 64–73°F (18–23°C). Some selected forms are grafted, but this is becoming increasingly rare.

Phyllodoce

These tiny erica relatives from the far north often self-layer and will also grow well from very small softwood and semi-ripe cuttings taken in late spring and summer. These are best struck under mist. The very fine seed should be sown, uncovered, on finely sieved sphagnum moss and must be kept cool and moist.

Physalis Ground Cherry

The ornamental species are usually propagated by dividing established clumps in early spring. Their seeds and those of the edible tomatillo (*P. ixocarpa*) can be extracted from the ripe fruit ready for sowing in spring. The seed germinates within two to three weeks if sown at around 68°F (20°C) with exposure to light.

Physostegia Obedient Plant

The common obedient plant (*P. virginiana*) is easily propagated by seed, division in late winter and early spring or by taking softwood cuttings of the non-flowering shoots during the growing season. Sow the seed, lightly covered, at around 68°F (20°C); seed is, of course, not a suitable method of propagation for the cultivars.

Phyteuma Rampion

These small perennials may be raised from seed or by division. The seed needs only a light covering and if sown in spring at around 68°F (20°C) the seedling should be up within three weeks. Established clumps may be divided in late winter or early spring just as the new growth commences.

Phytolacca Pokeweed, Pokeberry

This genus includes several species that can develop into trees but the plants most often seen in cultivation are vigorous perennials. They are usually raised from seed that can be sown fresh in fall or collected and stored until spring. Brief stratification may help but doesn't appear essential. The seed should be covered and kept at around 68°F (20°C).

Phytolacca rugosa

Picea Spruce

Species spruces are easily raised from seed collected from cones that have been allowed to dry naturally. Stratify the seed for about 12 weeks and then sow covered at around 64°F (18°C). Selected forms can be propagated by semi-ripe cuttings of side-shoots, although these can take up to two years to strike. Most of the best forms, usually selections of *P. pungens*, are grafted on to seedling stocks of the same species.

Picea mariana

Pieris Lily-of-the-Valley Shrub

These evergreen shrubs are very easily grown from seed sown at moderate temperatures of around 59–68°F (15–20°C) on a mix high in sphagnum moss. However, most garden plants are hybrids or cultivars that are propagated vegetatively, sometimes from layers but more commonly by taking softwood or semi-ripe cuttings from spring to autumn, preferably with mist and bottom heat.

Pieris japonica 'Valley Rose'

Pinus Pine

Pines are most easily propagated by seed. The exact germination requirements vary with the species, some need stratifying, but most germinate very well if sown, covered, at around 68°F (20°C). Relatively few pine hybrids or cultivars are seen in gardens, but those that are grown are usually propagated by summer semi-ripe cuttings or winter hardwood cuttings, which often strike better if taken with a small heel.

Pinus thunbergia

Pittosporum

The name pittosporum means sticky seed, and indeed their stickiness can make them hard to handle. So although they do not really need soaking to germinate, it helps remove the coating. Sow the seed at around 68°F (20°C) and cover it. The many cultivars, especially those of *P. tenuifolium*, strike well from semi-ripe summer to autumn cuttings.

Pittosporum crassifolium

Platanus Plane, Sycamore

Although seeds germinate quite freely if sown in spring after stratification, many garden planes are hybrids or cultivars that must be propagated vegetatively. Winter hardwood cuttings strike well and are usually ready for transplanting by the following winter. Any basal suckers may be pegged down as layers and aerial layering is also possible.

Bark of *Platanus orientalis*

Platycodon grandiflorus

Plectranthus saccatus

Pleione 'Versailles'

Plumbago auriculata 'Alba'

Plumeria rubra

Podalyria calyptrata

Platycodon Balloon Flower

This perennial is most commonly propagated by seed or division but may also be raised from its spring shoots by removing them and treating them as softwood cuttings. The seeds are best sown in autumn. They need exposure to light and germinate at around 64°F (18°C). Potted over winter they should be ready to plant out by mid-spring. Established plants may be divided in early spring but sometimes take a while to recover.

Plectranthus

This large genus of annuals, perennials and shrubs is easily propagated from seed, division or cuttings depending on the growth type. However, cuttings are by far the most common method for all but the annuals, which are rarely cultivated. Softwood and semi-ripe cuttings strike quickly during the warmer months and cuttings can be taken year-round if mist and bottom heat are available.

Pleione

These very showy hardy orchids are propagated by seed or division. As with most orchids the seed is very fine and germinating it is something for real orchid devotees. Division, however, is more straightforward, especially as *Pleione* pseudobulbs usually multiply well. They may be broken up in winter and early spring into clumps of a few pseudobulbs.

Plumbago Leadwort

The cultivated perennials and semi-climbing shrubs in this genus are usually propagated from cuttings. Warm season softwood and semi-ripe cuttings of non-flowering shoots strike well, especially with mist and bottom heat. Propagation from seed is also straightforward. Sow lightly covered at around 72°F (22°C) and the first seedlings should appear within three weeks.

Plumeria Frangipani

Renowned for the scent of its flowers, frangipani is usually propagated from 3–6 in (7.5–15 cm) long cuttings of side-shoots taken in late winter to early spring. These will be firm but not quite hard and still bare of leaves. The cuttings may take a considerable time to strike and must be kept warm and humid, so mist and bottom heat are definite advantages.

Podalyria Sweet Pea Bush

Propagated by seed or cuttings, the commonly grown species, *P. calyptrata*, is an easily cultivated shrub for mild gardens. The seed is freely produced and easily harvested. Soak it for 24 hours before sowing, covered, at around 72°F (22°C). Semi-ripe cuttings are best taken in summer and autumn and are fairly successful, though sometimes they seem to just sit around without forming roots.

Podocarpus Plum Pine

This genus of evergreen conifers has a wide distribution and includes many majestic trees. They may be raised from seed or cuttings. In most cases stratification helps improve the speed and evenness of germination but is probably not essential for the sub-tropical species. Cover the seed and keep at around 68°F (20°C); germination sometimes takes a considerable time. Take softwood and semi-ripe cuttings with a small heel at any time; they will strike better with a little bottom heat.

Podocarpus totara 'Pendula'

Podophyllum

Propagate these beautiful woodland perennials by sowing seed or dividing well-established clumps. The seed needs a period of cold: Sow outdoors in fall where the winters are cold or stratify for six weeks before sowing in spring in milder areas. It needs covering, germinates best at 59–68°F (15–20°C), and seedlings will start to appear after three weeks. Clumps may be divided in autumn or very early spring and usually re-establish quickly.

Podophyllum difforme

Podranea Port St. John Creeper

This super-vigorous climber may be propagated by seed and often self-sows but because cuttings yield flowering plants right away they are generally preferred to seed. Softwood and semi-ripe cuttings may be taken throughout the growing season and will quickly form roots. Any long, trailing stems may be layered.

Podranea ricasoliana

Polemonium Jacob's Ladder

Division when dormant is the usual way of increasing these hardy, easily grown perennials. They may, however, also be raised from seed, which should be sown as soon as it is ripe and kept at around 72°F (22°C), or by taking spring to early summer cuttings of the basal shoots.

Polemonium caeruleum

Polygala Milkwort

This large and variable genus includes annuals, perennials and shrubs. The commonly grown species are evergreen shrubs that are mostly propagated by taking semi-ripe cuttings. These strike well at any time but develop quickest during the warmer months. Winter hardwood cuttings are also fairly reliable.

Polygala chamaebuxus var *purpurea*

Populus Poplar, Aspen, Cottonwood

Despite the masses of downy seed frass of some species, poplars tend to produce relatively few seeds and because the seeds are small and have a short viability, most types are increased by taking open-ground 8–12 in (20–30 cm) long, winter hardwood cuttings of well-ripened one-year-old wood. These strike very freely and it is possible to start a poplar shelter hedge simply by inserting a row of cuttings.

Populus nigra var *betulifolia*

Potentilla cultivar

Potentilla Cinquefoil

Most of the 500 or so species are herbaceous perennials, though there are a few shrubs, too. Although the seed of some species needs stratification, it usually germinates freely at around 68°F (20°C). The clump-forming perennials can be propagated by division when dormant or by removing rooted runners or layers.

Softwood and semi-ripe cuttings of the shrubbier types and the woody perennials strike well at any time during the growing season.

Primula japonica hybrids

Primula Primrose

Primroses and polyanthus are easily raised from seed, but it must be fresh and must be kept cool at all times. The seed of some of the alpine species benefits from stratification. Don't cover the seed; simply sow it and keep it moist in a cool, shady location. It does not need more than about 59°F (15°C) to germinate. Once the seed is up, give it more light, but avoid exposing the tender seedlings to direct sunlight or prolonged heat.

Some primulas, such as the double-flowered polyanthus, can be divided, usually in late summer, just as the plants start into growth after their dormant period.

Prostanthera 'Ballerina'

Prostanthera Mint Bush

Most of these Australian evergreen shrubs are propagated by taking softwood or semi-ripe cuttings during the warmer months. The commonly grown types strike well but a few of the smaller species, such as *Prostanthera aspathaloides*, may be better raised from seed.

Protea cynaroides

Protea Sugarbush

Proteas are rarely easy to propagate. Grow the species from seed sown at 64–73°F (18–23°C) on a low-phosphate mix, but young seedlings are prone to damping off and low survival rates are the norm. The seedlings need good ventilation and just enough moisture to keep them growing.

Hybrids are usually propagated by semi-ripe cuttings taken from late summer, but be prepared for low strike rates and later collapse.

Prunus 'Ojochin'

Prunus Cherry, Plum, Almond, Apricot, etc.

Although flowering cherries are routinely budded or grafted, this is really more to produce trees of particular shapes, such as tall standards, than to simply perpetuate the variety. Most *Prunus* species, hybrids and cultivars will also grow from cuttings, either softwood in late spring and summer or hardwood during winter.

Plants may also be raised from seed, though with so many being cultivars this is rarely done unless grafting stocks are required.

Pseudopanax Lancewood

Several species of these evergreen shrubs and trees, mainly New Zealand natives, are known for having distinctly different foliage in their juvenile, adolescent and adult growth phases. To see these various phases it is necessary to raise plants from the seed, which is freely produced and easily germinated at around 68°F (20°C) provided it is fresh. Hybrids, cultivars and species that do not pass through a distinct juvenile phase are more often propagated by taking semi-ripe cuttings during the warmer months.

Pseudopanax 'Cyril Watson'

Ptelea Hop Tree, Water Ash

These interesting small North American deciduous trees are usually raised from seed, though selected forms are grafted onto seedling stocks. The seed should be stratified for 8–12 weeks before being sown, covered, at around 68°F (20°C). While the first seedlings should appear within three weeks, the seed can continue to germinate over quite a long period.

Ptelea trifoliata

Pterocarya Wingnut

These unusual trees are known for their conspicuous chain-like clusters of seeds, so not surprisingly seed is a common propagation method. Sow the seeds as soon as they are ripe and keep them at around 68°F (20°C). In areas with severe winters the seedlings may need protection for the first year. Hardwood cuttings and root cuttings taken in winter will strike but most often if vegetative propagation is necessary then the variety is grafted on seedling stocks of *P. fraxinifolia*.

Pterocarya x rehderiana

Pterostyrax Epaulette Tree

These deciduous trees, of which *P. hispida* is by far the most widely grown, are usually raised from seed or layers. Sow the seed in spring after stratifying for around eight weeks. Cover the seed and keep it at 64–75°F (18–24°C).

Pterostyrax hispida

Pulmonaria and *Omphalodes* Lungwort, Navelwort

Most garden forms of these spring-flowering perennials are hybrids or cultivars and should be divided during winter before they burst into growth. They will quickly re-establish with the arrival of warmer weather. Seed of the species germinates well if lightly covered and kept at around 64°F (18°C). Softwood cuttings of non-flowering stems may be taken from spring to early autumn.

Pulmonaria angustifolia

Pulsatilla Pasque Flower

Pasque flowers are beautiful and graceful spring-flowering perennials. The species are usually raised from seed, which should be lightly covered and kept at around 64°F (18°C). The cultivars, mostly forms of *P. vulgaris*, are propagated by dividing established clumps from autumn until late winter or very early spring. Root cuttings taken in winter strike well.

Pulsatilla vulgaris

Punica granatum

Puya berteroniana

Pyracantha 'Brilliant'

Pyrostegia venusta

Pyrus salicifolia 'Pendula'

Quercus robur

Punica Pomegranate

Pomegranates are, of course, absolutely full of seeds, and they are not difficult to geminate. After removing any fruit pulp, sow the seed, uncovered, at around 72°F (22°C). The best fruiting forms, such as 'Wonderful,' and the dwarf cultivars must be propagated vegetatively, either by taking softwood and semi-ripe cuttings in the warmer months or hardwood cuttings in winter. It is sometimes possible to remove rooted suckers.

Puya

These large, often spectacularly flowered bromeliads are most often increased from seed. Some species require brief stratification and most germinate at relatively cool temperatures of around 64°F (18°C). Established plants often develop small basal offsets that can be removed and grown on.

Pyracantha Firethorn

Most garden firethorns are hybrids or cultivars raised from softwood or semi-ripe summer cuttings. These strike well and are also an easy way to propagate the species. Later cuttings of firmer wood also strike but more slowly. If you want to try the seed, sow it fresh in the autumn and wait for the spring warmth to germinate it, or stratify it and sow it in spring. Cover the seed and keep it at around 64°F (18°C).

Pyrostegia Brazilian Flame Vine

The vivid flowers of the common species, *P. venusta*, are often seen brightening the winter and early spring days in mild climate gardens. It is an easily propagated plant that grows readily from softwood or semi-ripe cuttings taken during the warmer months and which may also be layered.

Pyrus Pear

The best fruiting forms of pears are grafted and sometimes double grafted to improve their disease resistance. Quinces (*Cydonia*) are often used as a rootstock but so is *Pyrus communis*, seedlings of which can be raised by sowing in spring seed that has been stratified for around 12 weeks. Cover the seed and keep at 59–72°F (15–22°C). Where blight is common, a disease-resistant form can be grafted on to the seedling stock and when the graft is set the fruiting variety can be grafted on to the blight-resistant wood.

Quercus Oak

Mighty oaks from little acorns grow. We all know that, so not surprisingly oaks are often raised from seed. It germinates best if sown at cool to moderate temperatures—around 64°F (18°C). Stratify the acorns of cool-temperate climate species before sowing. Oaks will also grow from cuttings, both summer semi-ripe and winter hardwood, although the strike rate varies markedly with the species.

Ranunculus Buttercup

This widespread genus includes annuals, biennials and perennials and their propagation varies accordingly. The annuals and biennials are raised from seed: sow, lightly covered, at around 61°F (16°C).

The perennials, which can be fibrous-rooted, stoloniferous or tuberous, are usually more easily propagated by division, which, depending on the dormant period of the species, can be done from mid-autumn to early spring.

Ranunculus asiaticus

Ratibida Mexican Hat

Very similar to *Rudbeckia* but far more drought tolerant, this genus of daisies may be propagated by seed or division. The seed germinates freely at around 68°F (20°C) and if sown in spring the new plants will be well-established by fall.

Large clumps can be divided in early spring as they start into growth and should continue to develop unchecked.

Ratabida columnifera

Rhaphiolepis Indian Hawthorn

Although these shrubs produce conspicuous small fruits, they are most often raised from cuttings. Softwood and semi-ripe cuttings strike well under mist and cuttings of firm one-year-old wood taken in spring will strike in outdoor beds, albeit often rather slowly.

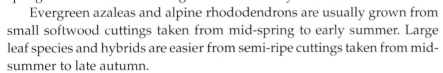

Rhaphiolepis umbellata

Rhododendron

This is a large genus with varying propagation requirements. Nearly all, including the azaleas, grow well from seed, which germinates freely at moderate temperatures of around 59–73°F (15–23°C) on a mix high in sphagnum moss. The seedlings will take a few years to flower well.

Evergreen azaleas and alpine rhododendrons are usually grown from small softwood cuttings taken from mid-spring to early summer. Large leaf species and hybrids are easier from semi-ripe cuttings taken from mid-summer to late autumn.

Deciduous azalea cuttings should be just on the verge of semi-ripe and taken in late spring. Some varieties do not strike well, if at all. Deciduous azalea cuttings must be well established by winter, or they will need to be kept indoors to prevent collapse. Mist and bottom heat make a considerable difference.

Rhododendron cultivars

Rhodohypoxis Rose Grass

These small, spring-flowering South African perennials quickly form clumps of foliage and are very easily propagated by division when dormant. This is best done in late winter or very early spring just before the new growth commences.

Rhodohypoxis baurii var *confecta*

Rhodotypos scandens

Rhodotypos

Related to the roses and with a flower that resembles a small, single, white rose, the sole species in this genus is easily raised from seed or cuttings. The seed should be sown at moderate temperatures in spring after 8–12 weeks of stratification. Alternatively, semi-ripe cuttings can be taken during the growing season or winter hardwood cuttings may be used. Layering is also very successful.

Robinia pseudoacacia 'Frisia'

Robinia Locust, False Acacia

These large deciduous trees produce conspicuous seedpods containing hard bean-like seeds. The seeds germinate well at around 64°F (18°C) and are best soaked and briefly stratified before being sown covered. Robinias do not strike very well from cuttings, but many develop suckers that can be lifted during winter sometimes with roots attached. The fancy foliage forms, such as the very popular 'Frisia,' are usually grafted on to stocks of the same species.

Rodgersia pinnata 'Elegans'

Rodgersia

These large perennials have strong rhizomes and form dense clumps of growth. The clumps may be divided in late winter or early spring and will quickly re-establish. Dividing rodgersias is, however, hard labour and it may be easier to sow seed. Sow outdoors in the autumn as soon as the seed is ripe, or stratify the seed in a refrigerator and sow it in spring. Kept lightly covered at around 64°F (18°C), the first seedlings should appear in a couple of weeks.

Romneya coulteri

Romneya California Tree Poppy

Somewhere between a perennial and a shrub, the California tree poppy is most easily raised from seed. Unless, that is, you already have a plant from which you can take root cuttings. Taken in the winter, root cuttings usually develop foliage quickly but can be slow to send out new roots. However, provided you start them in pots and can simply leave them to establish, this really is not a problem. Cover the seed lightly and keep at around 68°F (20°C).

Romulea bulbocodium

Romulea

Cormous perennials, these African and Mediterranean natives are mainly spring- to early summer-flowering and will soon form clumps of often rather grassy foliage that dies away during the dormant season when the plants can be divided. Alternatively, plants can be raised from spring-sown seed. Cover and keep at 64–75°F (18–24°C); it will germinate in three to six weeks.

Rosa Rose

The common hybrid tea and floribunda roses are usually budded on to selected vigorous rootstocks, which you may be able to obtain from a commercial grower if you wish to do your own budding. Species and miniature roses are most often propagated by softwood or semi-ripe cuttings and grown on their own roots. Old roses may be grown on their own roots or budded. Most rose species will grow well from seed, but it needs stratifying for up to 12 weeks before sowing. Try sowing fresh seed outdoors in autumn if you wish to avoid stratification.

Climbing roses

Rosmarinus Rosemary

Rosemary grows easily from seed or cuttings and the trailing forms often self-layer or you can induce layers to strike. Softwood and semi-ripe tip cuttings can be taken whenever they are available. They strike faster in warmer weather but winter wood is still reliable. Any branch wounded and pegged down to the soil will layer and can be removed for growing on.

Rosemary seed, while not suitable for propagating the cultivars, germinates freely and should be sown, uncovered, at around 68°F (20°C).

Rosmarinus officinalis var *prostrata*

Rubus Bramble, Raspberry, Blackberry, etc.

These vigorous cane-stemmed shrubs and climbers with their often-delicious fruits are most commonly increased by layering. In autumn the long, whippy canes are pegged down at their tips and should have developed a good root system by the end of the following growing season. Winter hardwood cuttings will also strike, as will summer semi-ripe cuttings. Some types have rootstocks that can be divided in winter. Plants may also be raised from seed but the best edible and ornamental types should be vegetatively propagated.

Rubus 'Benenden'

Rudbeckia Coneflower

Rudbeckias are perennials but some, like the popular *R. hirta*, are short-lived. These types are generally treated as annuals and raised afresh from seed each year. The seed should be sown at around 72°F (22°C), lightly covered.

The more reliably perennial types can be divided during winter or in very early spring, or their strong young shoots can be used as cuttings.

Rudbeckia fulgida var *deamii*

Russelia Coral Plant

This unusual shrub has somewhat weeping rush-like stems with sprays of pinkish red flowers. Sometimes the plants can be divided or layered, but most often they are propagated by taking cuttings of non-flowering side-shoots during the warmer months. These are best taken with a small heel and will strike much more readily with mild bottom heat.

Russelia equisetiformis

205

Salix sp.

Salvia involucrata

Sandersonia aurantiaca

Sansevieria trifasciata 'Britannica'

Santolina chamaecyparissus

Sarcococca ruscifolia

Salix Willow, Osier

Willow cuttings strike very easily. Just about any twig pushed into the ground over winter will leaf up in the spring. Alternatively, try layering: the whippy stems bend to ground level without too much of a fight and soon develop roots if wounded and pegged down. Willow cuttings are so simple to strike that the seed is seldom sown.

Salvia Sage

Annual and perennial salvias may be grown from spring-sown seed kept at around 70°F (21°C). Germination takes ten to 20 days. The cultivars and hybrids are propagated by dividing established clumps in late winter and early spring, from cuttings of the fleshy new shoots in spring, or by taking cuttings of non-flowering shoot tips at any time through the growing season.

Sandersonia Golden Lily-of-the Valley

This perennial with its pretty and interesting golden bell-like flowers will, if happy, produce many small tubers and propagation is just a matter of lifting the clump in late winter or early spring and breaking it up. The seed is often available and germinates well, albeit sometimes slowly. Cover lightly and keep at around 72°F (22°C).

Sansevieria Bowstring Hemp

Although the ability of these plants, popular outdoors in mild gardens and indoors elsewhere, to grow from lengths of leaf is well known, when propagated this way the variegations can become unstable. It is generally better to divide the rootstock of established clumps during the warmer months, if possible into pieces with reasonably large whorls of foliage.

Santolina Lavender Cotton

These small evergreen shrubs, popular for edging herb gardens and in large rockeries, are most often raised from small softwood and semi-ripe cuttings taken during the growing season, though cuttings will strike at any time with bottom heat. Plants may also be raised by sowing seeds in spring, lightly covered at around 68°F (20°C).

Sarcococca Sweet Box

Softwood and semi-ripe cuttings usually strike well but take a long time and although this can be sped up slightly with bottom heat, many propagators prefer to raise these plants from seed. Sow outdoors in the autumn for spring germination or stratify the seed for around eight weeks before sowing at 59–72°F (15–22°C). Kept covered, it will take at least three weeks to germinate. For vegetative propagation layering is often more practical than cuttings.

Saxifraga Saxifrage

There are around 370 species of saxifrage and at least as many hybrids and cultivars. Yet while being a diverse lot, they are all propagated in much the same way. However, they do not necessarily all react to propagation in the same way. Most eventually form clumps that can be broken up, either into smaller clusters of rosettes or tufts of foliage. Some recover quickly from this treatment, others don't.

The seed will usually germinate fairly well if sown at around 68°F (20°C), very lightly covered. The seedlings, however, can be tricky to handle. If practicable leave them as long as possible to develop good root systems before transplanting.

Saxifraga 'Triumph'

Schefflera Umbrella Tree

This large, principally tropical and sub-tropical, genus of trees, shrubs and a few vines includes several species that are popular in warm gardens or as houseplants. Although they produce large amounts of seed, which germinates quite freely at around 72°F (22°C), most cultivated plants are raised from semi-ripe cuttings taken during the warmer months.

Indoor umbrella trees are often aerial layered when they become too large and the rooted top section is used as a new, shorter plant.

Schefflera veitchii

Schinus Pepper Tree

The two commonly grown species in this genus are evergreen trees that while very ornamental can become invasive in warm areas. So not surprisingly they are easily raised from seed, which germinates within two to three weeks if kept at 64–77°F (18–25°C) and lightly covered. Semi-ripe cuttings strike quite well during the warmer months.

Schinus molle var *areira*

Sedum Stonecrop

This usually succulent genus of over 300 species includes annuals, perennials and sub-shrubs. They can be raised from seed, division and cuttings and are very easily propagated. The seed should not be covered and germinates best at warm temperatures, around 77°F (25°C).

Those that take root as they spread can be broken up at any time but re-establish most rapidly if divided in spring, while small pieces of stem soon strike roots if used as cuttings. Many species will also grow from leaf cuttings.

Sedum spurium 'Schorbuser Blut'

Sempervivum Houseleek

Propagating houseleeks is largely just a matter of breaking up the clusters of rosettes and treating them as individual plants. There is no reason why you cannot sow the seed, just cover it lightly and give it around 75°F (24°C), but there is hardly ever any need to bother, unless it is the only way you can obtain new or rare plants.

Sempervivum cultivars

Senna multiglandulosa

Sequoiadendron giganteum

Sidalcea malviflora

Silene uniflora

Sinningia speciosa cultivar

Senna

This genus now includes many of the shrubs and trees formerly listed under *Cassia*. All are notable for their flowers, which are often yellow, and their conspicuous seedpods, filled with hard brown or black bean-like seeds. Scarify the seeds and preferably soak them, too. Afterwards, sow them, covered, at around 75°F (24°C).

Softwood and semi-ripe cuttings can be taken from early summer to mid-autumn. The few perennials in the genus, such as *S. marilandica*, can be divided in spring, just as growth commences.

Sequoiadendron Giant Sequoia, Redwood

Even more extreme than mighty oaks from tiny acorns, the redwood is usually raised from the small seeds found within its cones. These germinate more evenly if stratified for around eight weeks and will start to appear in three to six weeks if covered and kept at 59–72°F (15–22°C). Selected forms, such as 'Pendulum,' are increased from semi-ripe to hardwood cuttings that can take a long time to strike.

Sidalcea Prairie Mallow

These upright perennials soon form sturdy clumps that can be divided in late winter or early spring. Seeds may also be sown and basal stems can be used as cuttings but division is so straightforward the other methods are seldom necessary.

Silene Campion, Catchfly

With over 500 species, this genus encompasses a huge range of annuals, biennials and perennials that are propagated by seed, division or cuttings. The seeds are very small and germinate quickly if sown, lightly covered, at around 68°F (20°C).

The clump-forming types may be divided when dormant from late autumn to early spring. Softwood cuttings of the spring stems or later cuttings of non-flowering stems strike well in the warmer months, especially with mild bottom heat.

Sinningia Gloxinia

Gloxinias can be rather tricky plants to grow—subject to mildew and low humidity problems—but they are not difficult to propagate. Established plants can be divided as they start into growth or small lengths of stem may be used as cuttings, but most growers prefer to take leaf cuttings and strike them in very humid enclosed conditions.

Plants can also be raised from the very fine seeds: sow, uncovered, on a peaty mix at 64–73°F (18–23°C).

Skimmia

Because *Skimmia* is a largely dioecious genus and many garden plants are cultivars grown for their berries, vegetative propagation is generally preferred to ensure the sex of the plants is known. Softwood and semi-ripe cuttings can be taken in the warmer months or hardwood cuttings may be used in winter. Softwood cuttings strike well under mist but otherwise the strike rate is rather variable.

Skimmia japonica

Solanum

The 1,400 species of the potato family cover a wide range of plant types. They are not all tuberous-rooted perennials, many are climbers, others are shrubs and there are even a few trees. The tuberous perennials, typified by the common edible potato, are easily propagated by division when dormant. The climbers grow most readily from semi-ripe summer cuttings and they also layer well. The shrubs and trees strike freely from semi-ripe cuttings. The seed of most species needs light to germinate and should not be covered. Sow it at around 68°F (20°C).

Solanum pseudocapsicum

Soleirolia Baby's Tears

This tiny-leaf, prostrate, evergreen perennial is a very popular small-scale groundcover in shaded parts of the garden and indoors for carpeting pots and terrariums. Although it often self sows, few gardeners collect the minute seeds, preferring instead to break off rooted pieces or take small lengths of stems for cuttings, both of which may be done with complete success at any time.

Soleirolia soleirolii 'Marginata'

Solidago Goldenrod

Goldenrods are extremely tough perennials that are late-flowering standards in the herbaceous border. They are very easily propagated by dividing established clumps in winter and very early spring when they are dormant or just starting into growth. They also grow well from seed, which should be stratified for eight weeks and sown in spring, lightly covered, at around 64°F (18°C).

Solidago canadensis 'Baby Gold'

Sophora Kowhai, Pagoda Tree

The seed of many species has a very hard coat and can survive prolonged exposure to saltwater, which is how the genus colonized many parts of the Pacific Rim. The seed must be sown very fresh or given prolonged soaking in warm water and then scarified before being sown, covered, at around 68°F (20°C).

Summer semi-ripe cuttings strike fairly well and sometimes do better if taken with a small heel. Those that are hard to strike from cuttings may be aerially layered. The weeping forms of *S. japonica* are grafted on to prepared standards of the same species raised from seed.

Sophora tetraptera

209

Sorbus aucuparia

Sphaeralcea ambigua

Spiraea japonica 'Nyewoods'

Sprekelia formosissima

Stachys byzantina

Stachyurus chinensis

Sorbus Rowan, Whitebeam

If your winter is cold enough to provide natural chilling, sow the seed outdoors as soon as it is ripe, otherwise stratify the seed before sowing. It needs quite a long chilling period—around 14 weeks. Although initially the seedlings may be very slow, they develop quickly after a couple of years.

Rowan cuttings seldom strike well, so the cultivars are usually raised by layering plants that have been coppiced to produce low branches.

Sphaeralcea Globe Mallow

These attractive and long-flowering perennials are tolerant of drought and neglect and are easily propagated by seed, cuttings or division. If sown in spring at around 72°F (22°C) the seeds will germinate within a few weeks.

Softwood and semi-ripe spring and summer cuttings are equally reliable and well-established clumps can be cut back hard in the fall so that they may be divided from late winter.

Spiraea

These hardy deciduous shrubs are easily propagated by taking softwood and semi-ripe cuttings in summer or hardwood cuttings in late autumn and winter. Many of the garden plants are hybrids or cultivars, so seed is not a suitable propagation method, but if you want to raise the species from seed, sow it outdoors fresh in autumn or stratify it and sow it in spring.

Sprekelia Jacobean Lily

This Central American bulb is a surprisingly tough plant that soon multiplies to form a sturdy clump of strappy leaves with dusky red flowers. Propagation is a simple matter of dividing the clump once the foliage dies off for winter.

Stachys Betony

Perennials related to mint, these plants are most often increased by division or by removing the many small offsets that form around the foliage clump. One commonly grown species, Chinese artichoke (*S. affinis*), has edible tuberous roots that can be lifted and stored over winter for replanting in spring. Plants can also be raised from uncovered seeds sown in spring at around 68°F (20°C).

Stachyurus

Not spectacular, but interesting, these late winter-flowering shrubs are usually raised from cuttings. Softwood and semi-ripe cuttings taken during the warmer months strike fairly well, especially with mist and bottom heat, but need to be well established before over-wintering outdoors.

Stephanotis Madagascar Jasmine

Popular as a cut flower for weddings, hence its alternate name of bridal wreath, this tender climber may be increased by seed or cuttings. The seeds take a while to germinate and need warm temperatures but are reliable provided they are fresh.

Softwood and semi-ripe cuttings taken during the growing season have a variable strike rate but are reasonably successful with mild bottom heat.

Stephanotis floribundum

Stewartia

These deciduous relatives of the camellias are grown for their flowers, autumn foliage and often interestingly colored and patterned bark. The usual propagation method is to take softwood or semi-ripe cuttings in spring and summer. These strike reasonably well but can take a long time.

Growing from seed is an alternative but although their seeds ripen in autumn many stewartias require a period of warmth followed by cold to break dormancy. Sowing the seed outdoors when ripe and waiting patiently may be the best method.

Stewartia pseudocamellia

Stokesia Stoke's Aster

The sole species in this genus is very easy to propagate. You can sow seed, divide it or take cuttings. Sow the seed, lightly covered, at 68°F (20°C). Divide established clumps in late winter or take softwood cuttings of the young shoots during the growing season. Whichever method you chose, you should have success.

Stokesia laevis

Strelitzia Bird of Paradise

Strelitzia seed is very hard and requires soaking for about three days before sowing. Even then, scarifying is recommended. Sow the seed with warm conditions, over 77°F (25°C), and lightly cover it. Because of these requirements it is fortunate that *Strelitzia* can also be propagated by division. However, they have very strong roots that are often difficult to prize apart. If you want to tackle the job, do it at any time from late autumn to early spring.

Strelitzia reginae

Streptocarpus Cape Primrose

While Cape primroses are easy enough to raise from seed—sow uncovered at around 75°F (24°C)—they can also be propagated by division and leaf cuttings. Division is only possible with large plants that have developed several foliage rosettes, and even then avoid making your divisions too small. Allow at least two rosettes per division. Leaf cuttings are the easiest way to produce several plants of one type and they can be done at any time during the warmer months.

Streptocarpus x hybridus

211

Styrax obassia

Styrax Snowbell

While these attractive spring-flowering deciduous shrubs and trees can be raised from summer cuttings or layers, seed is usually the best means of propagation. Sow the seed outdoors when ripe in autumn or stratify it in a refrigerator for six weeks before sowing, covered, in spring at 64–77°F (18–25°C). *Styrax* seed usually germinates well but occasionally requires two stratification periods. Leave ungerminated seeds outside over the following winter and they may sprout in the next spring.

Symphoricarpos albus

Symphoricarpos Snowberry

These deciduous North American shrubs are related to the honeysuckles(*Lonicera*) and bear similarly conspicuous berry-like drupes in unusual shades of pale pink to white. Although the drupes contain seeds, garden forms are most often propagated by open-ground hardwood cuttings. Take these as soon as the foliage has fallen; they should have struck and be safe to transplant by the following winter.

Syringa 'Belle de Nancy'

Syringa Lilac

Lilacs grow freely from seed, but it should be stratified for about two months. Sow it at moderate temperatures, around 64°F (18°C), in spring.

The hybrids and cultivars are usually grafted, often onto privet stocks, but they can just as easily be grown from layers or cuttings: semi-hardwood in summer to autumn, and hardwood in winter. The only drawback with lilacs on their own roots is that they sucker badly, but so does privet. The suckers are, of course, another source of propagation material.

Tamarix parviflora

Tamarix Tamarisk

Renowned for their ability to thrive in the worst soils and the harshest coastal winds, tamarisks are most easily propagated by taking hardwood cuttings in late autumn and winter. These can be struck in outdoor beds and transplanted during the following autumn. Semi-ripe summer cuttings also strike, though without mist they are generally less successfully. Sow the seed outdoors in autumn, lightly covered; it usually germinates well with the arrival of spring.

Taxodium distichum

Taxodium Swamp Cypress

Renowned for their ability to grow in water and for their unusual breathing roots or "knees," the swamp cypresses are deciduous conifers that are usually raised from seed. The seed does not need to be sown in water but does prefer damp soil. It germinates more evenly if stratified for 12 weeks before sowing and it needs covering and temperatures around 68°F (20°C).

Cultivars are propagated by taking semi-ripe summer cuttings which strike quickly in the warmth and high humidity of a mist propagation unit.

Taxus Yew

Yews have showy drupes containing a single seed, which usually germinates after passing through a bird and a winter. Other than removing the drupe flesh from the seed, there is no need to replicate the digestive process, but yew seeds do need around 12 weeks chilling. Afterwards, sow them, covered, and keep at around 68°F (20°C) and wait, because they can take a while to germinate. If that seems too complicated, take softwood or semi-ripe cuttings in summer or, even easier, take large tip cuttings in winter.

Taxus baccata 'Yew'

Tecomaria Cape Honeysuckle

This vigorous scrambler presents no propagation difficulties. Its seeds germinate well, but softwood and semi-ripe summer and autumn cuttings strike so freely that there is really no need to bother sowing seed.

Tecomaria capensis

Telopea Waratah

As with many protea family plants, waratahs are not easy to propagate, and that is only half the battle because once propagated they are difficult to keep alive and get established. It is partly to do with their close symbiosis with certain soil fungi, but is mainly just a trait of proteaceous plants in general.

Scarify the seed or soak before sowing and keep covered at around 68°F (20°C). Semi-ripe cuttings can be taken in summer to mid-autumn and sometimes strike well. Some of the newer cultivars have been propagated by tissue culture.

Telopea speciosissima

Teucrium Germander

This largely Mediterranean genus includes over 300 species of perennials, sub-shrubs and shrubs. They are propagated by seed or cuttings, with the perennials offering the additional option of division. Seed germinates at around 68°F (20°C): sow, lightly covered, in spring. Divide perennials during the cooler months and take semi-ripe cuttings of the shrubby types whenever material is available.

Teucrium cossonii 'Majoricum'

Thalictrum Meadow Rue

The meadow rues are graceful perennials that are usually propagated by dividing established clumps when they are dormant. The species can also be raised from seed: sow, lightly covered, at around 68°F (20°C).

Thalictrum minus

Thryptomene

These wiry-stemmed Australian evergreen shrubs may be grown from spring-sown seed kept at 64–75°F (18–24°C) but are more often propagated by taking semi-ripe cuttings in summer and autumn. A misting unit will allow for smaller cuttings that strike more quickly.

Thryptomene calcyina

Thuja standishii

Thymus cilicicus

Tiarella cultivar

Tibouchina granulosa 'Kathleen'

Tilia platyphyllos

Thuja Arborvitae

While these very hardy and adaptable conifers are quite easily raised from seed, most of the garden specimens are cultivars that must be propagated vegetatively. Use semi-ripe cuttings in summer or hardwood cuttings in winter. *Thuja* cuttings strike more reliably if taken with a small heel.

Thymus Thyme

Thymes are easily propagated by several methods. The species are easily raised from seed and often self-sow in profusion, coming up in every little nook and cranny. For more control, sow the seed in trays, lightly covered, at around 68°F (20°C).

The carpeting types often strike roots as they spread and small rooted pieces can be broken off and grown on. Propagate those with woody stems by layers and by taking small tip cuttings at any time during the growing season.

Tiarella Foamflower

Related to *Heuchera*, and sometimes crossed with that genus to produce the intergeneric hybrid x *Heucherella*, these low-growing, spring- to early-summer-flowering perennials are most often propagated by dividing well-established clumps as they start into growth in late winter. They may also be raised from seed sown in spring at around 68°F (20°C).

Tibouchina Lasiandra, Glory Bush

The velvety foliage and stunning flowers of these sub-tropical and tropical shrubs never fail to impress. Late spring to early autumn semi-ripe cuttings strike well and are even more successful with mist and bottom heat.

Tilia Lime, Linden

Limes are often grown from stratified seed sown at around 64°F (18°C) but the garden forms are more often propagated vegetatively. This can be done by layering or by taking soft to semi-ripe cuttings in summer. However, most commercial growers choose to graft cultivars on to seedling stocks in late summer. Growth compatibility between stock and scion is critical with *Tilia*.

Tolmiea Piggyback Plant

While popular as a houseplant for dark, draughty corners, this low, evergreen perennial is something of a weed in cool, moist climates. It self-sows readily but is best known for its habit of developing young plantlets on the back of its mature leaves. Leaves with plantlets can be detached and potted; they will soon form roots. Large clumps may also be divided.

214

Toona Toon

These deciduous trees, best known for the spring foliage of *Toona sinensis*, may be propagated from seed but, as the foliage color of seedlings is variable and cuttings largely unsuccessful, the preferred methods are to remove rooted suckers in the fall or to take root cuttings over winter.

Toona sinensis

Townsendia

Tough silver-leafed daisies from North America, these sometimes short-lived perennials can be divided with care but are more often raised from seed. Sow it in spring at around 20°C, lightly covered, and the seedlings should begin to appear within a couple of weeks.

Trachelospermum

The only commonly grown species of this genus, *T. jasminoides*, is an evergreen climber most often represented in cultivation by various selected forms. Seed is not a practical method of propagation for these cultivars as they will not reproduce true to type. Instead take softwood semi-ripe cuttings during the warmer months. These strike fairly well but often slowly. Layering is effective if only a few new plants are required.

Trachelospermum jasminoides

Tradescantia Spiderwort

This genus includes some 70 species, mostly perennials with a few annuals. They may be increased by seed, division or cuttings depending on the season and the growth type. The seeds should be lightly covered and will germinate at around 68°F (20°C). Species that form clumps that shoot from the base may be divided when dormant or at any time if well watered. Tip cuttings of non-flowering stems strike well; take whenever they are available.

Tradescantia virginiana 'Iris Pritchard'

Tricyrtis Toad Lily

Valuable for their autumn flowers and graceful growth habit, these hardy herbaceous perennials form sturdy clumps that can be divided when dormant in winter and very early spring. Most species grow quickly enough that they can be divided every two to three years.

Tricyrtis hirta 'Miyazaki'

Trillium Wakerobin, Trinity Flower

Although trilliums are often expensive plants to buy, they are not that difficult to propagate, at least not in small numbers. Established clumps can be broken up during winter and will quickly recover in the spring. It is often more practical to raise the smaller species from seed: stratify for ten weeks and sow at around 64°F (18°C). Because some trilliums require two periods of stratification, it is often easier, in areas with winters cold enough, to autumn-sow them in trays outdoors and wait for the seedlings to appear.

Trillium rivale

215

Trollius europaeus

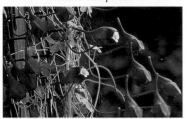

Tropaeolum tricolorum

Tsuga heterophylla

Tulipa cultivars

Ulmus procera 'Louis van Houtte'

Umbellularia californica

Trollius Globe Flower

Many of the garden specimens of these hardy buttercup relatives are cultivars or hybrids and must be propagated vegetatively. Divide the clump in winter or very early spring. The species may be raised from seed but most often they are divided, too.

Tropaeolum Nasturtium

Many of the *Tropaeolum* species develop masses of small tubers and propagation requires nothing more than lifting the roots when the plant is dormant, breaking up the tubers and replanting. Non-tuberous perennials are usually struck from late spring to mid-autumn softwood or semi-ripe cuttings and the annuals are raised from seed sown, covered, at around 68°F (20°C).

Tsuga Hemlock

These trees may be increased by sowing seeds outside in the autumn or in spring after stratification for eight weeks. They germinate at around 68°F (20°C) and should be covered. Many garden plants are cultivars, however: propagate vegetatively from semi-ripe cuttings taken in late summer and autumn. Seedlings are only really required as stock plants for growers who like to make grafts.

Tulipa Tulip

Tulips multiply reasonably well if left undisturbed. That is fine in cool climates but warm conditions after flowering can cause the bulbs to deteriorate and fail to produce offsets. Consequently, seed is often a more reliable method and as viruses transmitted from parent bulbs to the offsets can be a problem, it is an advantage that seedling tulips are also virus-free. Stratify the seed for at least eight weeks and sow, covered, at 59–64°F (15–18°C). Seedlings can take five years to bloom properly, so don't be in a hurry.

Ulmus Elm

Elms can be grown from softwood and semi-ripe summer cuttings, two-year-old hardwood summer cuttings, winter hardwood cuttings that are kept indoors with heat or winter hardwood cuttings kept outdoors. They may also be grafted, usually on to *U. glabra* stocks. If you want to raise seedlings for grafting stock, stratify the seed for 12 weeks and sow it in spring, covered, at around 64°F (18°C).

Umbellularia California Laurel

This western North American evergreen tree has very aromatic foliage that for some people causes headaches. So although it can be propagated from semi-ripe cuttings in summer and autumn it may be advisable to not handle the plant too closely and to raise seedlings instead. Sow the seed fresh, cover lightly and keep at around 68°F (20°C).

216

Vaccinium Blueberry, Cranberry, etc.

Encompassing around 450 species of shrubs, small trees and a few vines, this genus includes both evergreen and deciduous types. Small evergreen ornamental species, such as *V. delavayi*, are propagated from semi-ripe cuttings in summer and autumn and prostrate species, such as *V. macrocarpon*, that strike roots as they spread, can be divided or have rooted pieces removed. Deciduous types tend to be more difficult. Raise them from hardwood cuttings 4–6 in (10–15 cm) long, taken in spring as growth commences. Keep them cool, moist and shaded and plant out the following spring. Seedlings are seldom required but the seed will germinate quite readily if sown in spring, uncovered, on a damp peaty mix.

Vaccinium delavayi

Verbascum Mullein

This genus is very easily propagated by seed for the biennials, and seed or division of the perennials. The seed often self-sows in autumn but more controlled results are achieved by collecting it, then sowing in spring, uncovered, at 68–75°F (20–24°C). The perennial clumps are best divided in very early spring; some perennial species will also grow from winter root cuttings and can have their young spring shoots removed and treated as softwood cuttings.

Verbascum 'Helen Johnson'

Verbena

Annual bedding verbenas are raised from seed, lightly covered, with quite warm conditions, around 70°F (21°C). The perennial verbenas tend to be short-lived but are very easily propagated. They usually self-layer and small pieces can be broken off and grown on. Failing that, tip cuttings strike quickly and can be taken throughout the warmer months.

Verbena x *hybrida*

Veronica Speedwell

This genus of around 250 species of annuals and perennials, some rather shrubby, can be increased by seed, division or cuttings. The seed usually germinates in one to two weeks if sown in spring, uncovered, at around 72°F (22°C). Divide the clump-forming types when at their most dormant, from late autumn to early spring; those with woodier stems strike readily from warm-season softwood cuttings of non-flowering stems.

Veronica prostrata 'Mrs Holt'

Viburnum

Viburnums are vigorous shrubs that are not difficult to propagate. Summer softwood and semi-ripe cuttings strike reliably, as do hardwood winter cuttings, though things take a little longer in the winter. Most species come from areas with cold winters and the seed needs stratifying before sowing in spring. If it does not germinate well, it may need another period of chilling. Provided your winter is cold enough, it is often easier to sow the seed outdoors in autumn and just wait for nature to take its course.

Viburnum trilobum

Vitis vinifera cultivar

Weigela florida 'Variegata'

Wisteria floribunda

Yucca gloriosa 'Variegata'

Zantedeschia aethiopica

Zinnia elegans cultivars

Vitis Grape

While grapes may be raised from seed, almost all cultivated types, edible and ornamental, are cultivars or hybrids that need vegetative propagation. Softwood and semi-ripe cuttings strike freely in the warmer months and hardwood cuttings may be taken from late autumn. Layering works well, too. Commercial varieties are often grafted or budded on to stocks resistant to phylloxera and other pests and diseases.

Weigela

The seed of the species is best sown in spring at around 70°F (21°C). It takes at least 20 days to germinate and benefits from stratification. Hybrids and cultivars are best grown from cuttings, which strike very easily. Most methods work. Try softwood cuttings from late spring, semi-ripe in autumn, hardwood stem cuttings outdoors over winter, or hardwood tips indoors over winter.

Wisteria

Wisteria seed can be difficult to germinate. Soak the seed for 24 hours before sowing, and if the seed coats are still firm, scarify them. Sow the seed at cool to moderate temperatures of 57–64° F (14–18°C); plants will take from 30 to 50 days to show. Hybrid wisterias can be grown from cuttings, most commonly winter hardwood cuttings, or they may be grafted, usually on to seedling wisterias, but sometimes on to *Laburnum* stock.

Yucca

Low, clump-forming yuccas are sometimes divisible, if you are careful, but the most common method for these and taller-trunked species is to remove rooted offsets or suckers from the base. Offsets without roots can often be induced to form roots by treating them as cuttings. Some species, notably *Y. elephantipes*, will grow from lengths of stem with a leaf bud. These "logs" are inserted in pots and treated like large cuttings. The seed of most species will germinate well if sown lightly covered at around 68°F (20°C).

Zantedeschia Arum Lily, Calla Lily

Quite distinct from the true arums, these southern African lilies are easily propagated by seed or division. Divide large clumps in early spring; while the divisions can be unwieldy, they should grow well enough. The rhizomes of the species and hybrids with a clearly defined dormant period can be broken up when inactive. Sow fresh seed, uncovered, at around 72°F (22°C).

Zinnia

While there are a few perennial species, the common zinnias are annuals that are often required in large numbers for bedding displays. Propagate them by sowing lightly covered seed from late winter at around 72°F (22°C). The stems of the young seedlings are easily bruised, so take care.

Glossary

Adventitious A bud or root that appears on tissue where it would not normally be expected, e.g. aerial roots on *Hedera* (ivy) stems.

Aerial roots A root appearing above soil level, often from a branch. Used for both support and feeding, e.g. many orchids and *Monstera*.

Annual A plant that completes its life cycle from seed to maturity in one year.

Anther The pollen-bearing sac at the tip of a stamen.

Asexual propagation To produce plants without using seeds by vegetative methods such as cuttings, layering, tissue culture, etc.

Axil The upper angle between a leaf or stem and the branch from which it appears. Site of many shoots and flower buds, hence axillary.

Basal At the bottom, e.g. a basal shoot that appears where the stem emerges from the ground.

Base plate The flattened or conical stem within a bulb. Usually represented externally by a fleshy plate on the base of a bulb or corm.

Biennial A plant that completes its life cycle over two-years; grows and sets up reserves the first year, flowers, seeds and dies the next year.

Bisexual Having organs of both sexes functioning in the same flower.

Bottom heat Artificially heating the root zone of a cutting bed or pot, usually by electric heating pads or cables.

Bulblets Seedling bulbs or the small immature bulbs that form around a parent bulb during the growing season. Also known as bulbils.

Callus 1. The protective tissue that forms over a wound. 2. The white tissue that forms on the subterranean part of a cutting prior to root formation.

Cambium Basic growth cells. May be found in leaves, roots or stems but best known as the layer of growing cells immediately beneath the bark or skin of a stem.

Cane 1. The jointed stem of large grassy plants, e.g. bamboo. 2. The long arching growths of many plant genera, e.g. orchids, raspberries, roses.

Capsule A dry, divisible fruit composed of two or more sections.

Carpel One of the units comprising a pistil or ovary.

Clone An exact replica of the parent plant. Any plant propagated by vegetative means, such as division, budding, cuttings, layers. These methods are widely used for plants that do not grow true from seed.

Cone A dense construction of seed-bearing scales on a central axis, often woody and elongated (pine cones).

Corm An underground storage organ similar to a bulb but lacking scales, e.g. *Gladiolus.*

Cormlets Small seedling corms or the small immature corms that often form around a larger parent corm during the growing season. Also known as cormels.

Cotyledon The first leaf to emerge from a germinating seed. The fleshy seed leaves of a dicotyledon or the initial leaf of a monocotyledon.

Crown 1. The corona. 2. The base of a plant where stem and root meet. 3. Part of a rhizome with a bud, suitable for propagation by dividing, e.g. rhubarb

Cultivar A term for a variety that is maintained in cultivation as distinct from a variety that occurs naturally in the wild. A contraction of the term cultivated variety.

Cutting An amputated section of a plant or tree that will develop new roots and become self-sufficient. These may be taken from stems, branches, sometimes roots and leaves.

Dicotyledon A plant that produces two seed leaves. Usually the offspring of angiosperms.

Dieback A variety of fungal diseases that kill part or all of a plant by causing the tissues to die back from a tip or cut branch. Often due to faulty pruning techniques.

Dioecious With unisexual flowers, male and female blossoms borne on separate plants.

Divide To separate a clump of perennial plants into smaller clumps, hence division.

Dormancy The time when a plant makes minimum growth, Usually but not always occurring during winter. Often when a plant is bare of foliage. Hence dormant.

Embryo The young plant within the seed.

Endosperm The food storage tissue within a seed. Absorbed by the embryo during germination.

Eye 1. An undeveloped growth bud of a storage organ such as a bulb, corm or tuber. 2. A contrasting color spot in the center of a flower.

F1, F2 hybrid Respectively, the first and second generation offspring from a given parent plant or cross.

Family A group of related botanical genera.

Filament A thread-like organ, especially of a stamen supporting an anther.

Genus A grouping of closely related species.

Grex A group of seedlings from the same cross. A term most commonly used when referring to *Rhododendron* cultivars.

Harden off To acclimatize or prepare for colder weather.

Hardwood cutting A cutting taken from wood of the previous season's growth.

Heel A small strip of material from the main stem that is left on a side-shoot cutting. Usually obtained by tearing the side-shoot from the branch rather than cutting it. May improve the chances of a strike, particularly with conifers.

Herbaceous perennial A non-woody plant that dies back to the roots in winter, sending up new growth in spring.

Hormone A term used for many synthetically produced growth promoting or modifying agents, especially indolebutyric acid, a root-forming hormone.

Hybrid The result of cross-fertilization of different kinds of parent plants.

Inflorescence 1. The flowering part of a plant, irrespective of arrangement. 2. The arrangement of blooms in a flower head.

Internode The length of stem between two nodes, hence internodal and internodal distance.

Juvenile A young or immature plant. Many plants display distinct differences between juvenile and adult foliage and growth habit.

Lateral On or at the side, e.g. a side-branch produced from a main stalk or trunk.

Leader A plant's dominant shoot.

Leaf stage Growth stage of a crop or weed, usually expressed in number of leaves, e.g. second leaf stage.

Mature 1. A plant that has reached flowering age and is able to reproduce sexually. 2. A fully ripe fruit or seed.

Monocotyledon Plants that produce a single seedling leaf and includes grasses and cereals.

Mutant The result of a mutation. A variant, differing genetically and often visibly from its parent or parents and arising spontaneously.

Natural cross A hybrid that occurs without human help.

Node A point on a stem from which leaves develop.

Offset A small division from the side of a mature clump-forming plant. Usually able to be broken off without lifting the parent plant unlike division.

Ovary The structure at the base of the pistil in which the seeds of angiosperms develop.

Perennial A plant that lives for more than two years and regrows from the same stem or root system each year.

Petiole The stalk of a leaf.

pH The degree of acidity or alkalinity in soil based on the potential of hydrogen ions and measured on a logarithmic scale of from 0 (acid) to 14 (alkaline), with 7 as the neutral point.

Photosynthesis The process whereby plants use solar energy through the catalytic action of chlorophyll to convert water and carbon dioxide into carbohydrates.

Pollen The spores or grains borne by an anther, containing the fertilizing male gametes.

Pollination Applying pollen to the stigma.

Prothallus The sexual regeneration stage of a fern.

Pseudobulb The thickened bulb-like storage organ found in many orchids.

Rhizome An underground or surface creeping stem or root that enables plants to spread. May also act as a storage organ.

Rootstock A rooted section of plant used as the base onto which a scion from another plant is grafted.

Rosette An arrangement of leaves radiating from a crown or centre, usually close to the ground, e.g. Iceland poppies, lettuce and dandelion.

Scale A segment of bulb that may in some cases, e.g. lilies, be detached for propagation.

Scarify To weaken the covering of some hard-cased seeds to hasten germination. Large seeds can be nicked with a knife, smaller seeds rubbed with coarse sandpaper.

Scion A bud or shoot grafted onto the stock of another plant.

Softwood Unripened, immature tissue of any woody plant. Used for propagation in some species.

Spore The reproductive cell of ferns, mosses and fungi. Although distributed like seed, a spore differs from a flower's seed in that it is unfertilized at the time of dispersal.

Sport A mutation showing distinct variations from the norm, e.g. a different foliage form or flower color.

Stamen The pollen-bearing or male organ of a flower. Usually composed of a filament and an anther.

Stigma A sticky pad at the end of the style that receives the pollen.

Stock The parent plant onto which the scion or cutting is grafted. Also a term for the parent plant from which cuttings are collected.

Stolon A shoot that runs along the ground, taking root at intervals and giving rise to new plants, e.g. couch or kikuyu grass. An often misused term.

Stratify To treat dormant seeds by chilling under moist conditions to simulate winter conditions and effect germination—hence stratified.

Strike To cause a cutting to take root.

Style The tubular part of the pistil between the ovary and stigma, often elongated.

Sucker An adventitious stem arising from the roots of a woody plant, often from the stock rather than the scion of a grafted plant.

Systemic Any substance capable of permeating through the entire plant. Often said of insecticides and fungicides that destroy pests and diseases by circulating through the sap system.

Tender Intolerant of freezing conditions or prolonged exposure to low temperatures above freezing.

Thinning out 1. Pruning to reduce foliage density. 2. Removing some seedlings to allow more room for the remainder to grow.

Tuber A modified root that acts as a storage organ. Similar to a rhizome but usually shorter and thicker and does not elongate greatly as it grows, e.g. potatoes.

Tuberous root An underground storage organ that resembles a tuber but is actually a root. Growth buds form from the point where the previous year's growth stems were rather than from eyes, e.g. *Dahlia*.

Union The join between a rootstock and scion.

Unisexual Flowers with either functioning male or female organs but not both.

Varietal name A name used to describe a selected form as opposed to a hybrid, e.g. *Podocarpus totara* 'Aurea'. Often used interchangeably with cultivar.

Variety (officially Varietas) 1. A subdivision of a species. 2. A recognizably different member of a plant species capable of cultivation.

Vegetative 1. Those parts of the plant that are not flowering or involved with flowering. 2. A condition of growth in which flowering has not and is not about to occur.

Viability 1. The number of seeds in a group that are alive at any particular time. 2. The period of time after ripening in which a seed remains able to germinate. Proper storage can greatly extend viability but some plants, e.g. *Delphinium*, have a naturally short viability period.

Virus A minute organism, only able to replicate within a living cell, that causes discoloration, malformation or death. Few plant viruses have practical cures.

Index

Index of plant propagation

Index of common names